KNOWLEDGE MANAGEMENT FOR EDUCATIONAL INNOVATION

IFIP – The International Federation for Information Processing

IFIP was founded in 1960 under the auspices of UNESCO, following the First World Computer Congress held in Paris the previous year. An umbrella organization for societies working in information processing, IFIP's aim is two-fold: to support information processing within its member countries and to encourage technology transfer to developing nations. As its mission statement clearly states,

> *IFIP's mission is to be the leading, truly international, apolitical organization which encourages and assists in the development, exploitation and application of information technology for the benefit of all people.*

IFIP is a non-profitmaking organization, run almost solely by 2500 volunteers. It operates through a number of technical committees, which organize events and publications. IFIP's events range from an international congress to local seminars, but the most important are:

- The IFIP World Computer Congress, held every second year;
- Open conferences;
- Working conferences.

The flagship event is the IFIP World Computer Congress, at which both invited and contributed papers are presented. Contributed papers are rigorously refereed and the rejection rate is high.

As with the Congress, participation in the open conferences is open to all and papers may be invited or submitted. Again, submitted papers are stringently refereed.

The working conferences are structured differently. They are usually run by a working group and attendance is small and by invitation only. Their purpose is to create an atmosphere conducive to innovation and development. Refereeing is less rigorous and papers are subjected to extensive group discussion.

Publications arising from IFIP events vary. The papers presented at the IFIP World Computer Congress and at open conferences are published as conference proceedings, while the results of the working conferences are often published as collections of selected and edited papers.

Any national society whose primary activity is in information may apply to become a full member of IFIP, although full membership is restricted to one society per country. Full members are entitled to vote at the annual General Assembly, National societies preferring a less committed involvement may apply for associate or corresponding membership. Associate members enjoy the same benefits as full members, but without voting rights. Corresponding members are not represented in IFIP bodies. Affiliated membership is open to non-national societies, and individual and honorary membership schemes are also offered.

KNOWLEDGE MANAGEMENT FOR EDUCATIONAL INNOVATION

IFIP WG 3.7 7th Conference on Information Technology in Educational Management (ITEM), Hamamatsu, Japan, July 23-26, 2006

Edited by

Arthur Tatnall
Victoria University, Australia

Toshio Okamoto
University of Electro-Communications, Japan

Adrie Visscher
University of Twente, The Netherlands

Knowledge Management for Educational Innovation

Edited by A. Tatnall, T. Okamoto, and A. Visscher

p. cm. (IFIP International Federation for Information Processing, a Springer Series in Computer Science)

ISSN: 1571-5736 / 1861-2288 (Internet)

Printed on acid-free paper

ISBN 978-1-4419-4342-2 eISBN: 13: 978-0-387-69312-5

Copyright © 2007 by International Federation for Information Processing.
Softcover reprint of the hardcover 1st edition 2007
All rights reserved. This work may not be translated or copied in whole or in part without the written permission of the publisher (Springer Science+Business Media, LLC, 233 Spring Street, New York, NY 10013, USA), except for brief excerpts in connection with reviews or scholarly analysis. Use in connection with any form of information storage and retrieval, electronic adaptation, computer software, or by similar or dissimilar methodology now known or hereafter developed is forbidden.
The use in this publication of trade names, trademarks, service marks and similar terms, even if they are not identified as such, is not to be taken as an expression of opinion as to whether or not they are subject to proprietary rights.

9 8 7 6 5 4 3 2 1
springer.com

Table of Contents

Preface

As the editors of this volume we are very happy to publish a selection of the papers that were presented at the seventh Conference of Working Group 3.7 of the International Federation for Information Processing. The focus of Working Group 3.7 is on ITEM: Information Technology in Educational Management (for more information, please visit http://item.wceruw.org/), and the theme of its 2006 conference was on *Knowledge Management for Educational Innovation*. The event took place in Hamamatsu (Japan) and enabled the exchange of findings and ideas between researchers in educational management and information technology, policy-makers in the field of education, developers of ITEM systems, and vendors. The overall goal of the conferences of Working Group 3.7 is to demonstrate and explore directions for developing and improving all types of educational institutions through ITEM.

Contributions tot the conference came from all over the world: Spain, India, Australia, New Zealand, Taiwan, Hungary, England, Germany, Botswana, Japan, Uganda, and The Netherlands, and the number of papers was large (over 30). All papers in this book have been peer reviewed. They were selected from those presented at the conference and the authors given an opportunity to improve them before publication.

Contributions to the conference varied from innovative examples of how ITEM can support and improve educational practice at the level of instruction (e.g. in classrooms) and at institutional level, an analysis of the history of ITEM as a field of study and practice, results of research on how training can promote the implementation of ITEM systems, and theoretical analyses of the conditions under which ITEM will have the strongest impact. Moreover, all sectors of educational systems (from schools to universities) are represented by the various chapters in this book.

We hope the reader will enjoy the content of this book and that it will stimulate them in realizing that ITEM is a field where much is going on and where much can be gained if the potential of ITEM is utilized under the right conditions.

Last but not least, the reader is invited to one of our future conferences. The next one will be held in Darwin (Australia) in 2008; for more information, please have a look at our website in 2007.

Arthur Tatnall (Victoria University, Australia)

Toshio Okamoto (University of Electro-Communications, Japan)

Adrie Visscher (University of Twente, The Netherlands)

Ten Years of ITEM Research

Analysis of WG 3.7's Published Work (1994 – 2004)

Javier Osorio and Jacques Bulchand
Las Palmas de Gran Canaria University, Spain

Abstract: In this paper we review articles published in the proceedings of the conferences of the IFIP Working Group 3.7 in order to classify them by the most relevant topics and to identify the main research patterns followed by the Group. We have also established the major research methodologies used by the Group to carry out its work. Finally, we end up with a brief comparison with the information systems and technologies (IS/IT) field of study with the purpose, on the one hand, of recognising common tendencies and, on the other, of diagnosing how mature is information technology for educational management (ITEM) research.

Keywords: ITEM research, conference proceedings, research topics, research methodologies.

1. INTRODUCTION

After the first international conference on information technology for educational management (ITEM) took place in Jerusalem in 1994, Working Group 3.7 of the International Federation for Information Processing (IFIP) had held a total of five international meetings. At such reunions researchers, academicians and professionals have been able to contribute to a better development and understanding of all problems associated with the use of information technology (IT) for educational management. Most of these contributions, generally presented as papers, have been published and over time have become an important accrual of knowledge and experience.

ITEM studies have been considerably enriched by the varied profiles and backgrounds of the participants in these working conferences, to the point of acquiring certain characteristics of its own. This has already been put forward in *'ITEM: synthesis of experience, research and future perspectives on computer-assisted school information systems'* (Visscher *et al.*, 2001). Nevertheless, ITEM analysis has been approached from very different

Please use the following format when citing this chapter:

Osorio, J. and Bulchand, J., 2007, in IFIP International Federation for Information Processing, Volume 230, Knowledge Management for Educational Innovation, eds. Tatnall, A., Okamoto, T., Visscher, A., (Boston: Springer), pp. 1–8.

perspectives that respond to diverse academic disciplines, which is probably the underlying reason for the area's conceptual richness. After ten years of uninterrupted periodic meetings with outstanding academic results, it seems convenient to take a pause to reflect on what has been achieved in the past and to decide upon courses of action for the future. This is a necessary practice in any relatively new area such as this one, which is encumbered with certain difficulties owing to its immaturity and ambiguity, as well as its eclectic and interdisciplinary nature and the ever changing IT scenario.

Following this the paper's aim is to study the literature resulting from the international ITEM conferences organised by IFIP Working Group 3.7 in order to identify and classify the most researched topics or themes and the methodologies employed.

2. ANALYSIS METHODOLOGY

This study has been carried out on the basis of the publications that have followed the working conferences, which have compiled a selection of papers presented at the work sessions. These publications have been edited by prestigious companies specialised in publishing scientific papers. Table 1 shows the city where the conference took place, the year, title of the book, publisher, place of edition, date of edition and number of articles published in the book. In order to classify each article under a certain topic and methodology, we have studied the titles and abstracts, only reviewing the actual text in those cases in which classification was unclear according to the mentioned criterion.

Table 1: International ITEM conferences and resulting publications

Place	Year	Title of the book	Publisher	City	Year	Number of papers published
Jerusalem	1994	Information Technology in Educational Management	Chapman & Hall	London	1995	31
Hong-Kong	1996	Information Technology in Educational Management for the Schools of the Future	Chapman & Hall	London	1997	26
Maine	1998	The Integration of Information for Educational Management	Felicity Press	Maine	1998	17
Auckland	2000	Pathways to Institutional Improvement with Information Technology in Educational Management	Kluwer	Boston	2001	11
Helsinki	2002	Management of Education in the Information Age: The Role of IT	Kluwer	Boston	2003	14
Gran Canaria	2004	Information Technology and Educational Management in the Knowledge Society	Springer	New York	2005	18

The resulting information has been classified into three groups. Firstly, according to topics studied in each article. Secondly, considering the research methodology applied to each paper to further group them considering the type of approach employed. Finally, we have compared results with a similar analysis carried out by Claver *et al.* (1999) applied to IS/IT research, as there are solid common elements in both fields.

3. DATA ANALYSIS

Table 2 contains the most relevant topics discussed at the international conferences of IFIP Working Group 3.7 as they have been published. The table shows the number of papers that have dealt with each topic and the percentage these represent of the total number of papers both of each publication as well as the accumulated total, which is shown in the last column. We have decided to use large work areas for our classification in an attempt to obtain significant results. Otherwise, a more detailed classification might have made the results confusing or unclear. For example, Grover *et al.* (1993) suggests 20 IT related topics applicable to any study area, and authors such as Claver *et al.* increase this number up to 30.

Table 2: Papers classified by research subject

Topic	CONFERENCE						
	1994	1996	1998	2000	2002	2004	Total
	N. %	N. %	N. %	N. %	N. %	N. %	N. %
Strategies to integrate IT into educational management	2 6.5	1 3.8	2 11.8	1 9.1	1 7.2	2 11	9 7.7
Assimilation and integration of IT into educational management	4 12.9	9 34.6	6 35.2	3 27.3	4 28.6	1 5.5	27 23
ITEM state of the art. The discipline's present situation and trends	1 3.2	1 3.8	- -	- -	- -	1 5.5	3 2.7
Assessment of IT support to educational management	2 6.5	5 19.2	2 11.8	4 6.3	2 14.3	5 27.9	20 17.1
National, regional and local experience in the use of IT for educational management	10 32.3	- -	2 11.8	1 9.1	1 7.1	4 22.3	18 15.4
IT applications in educational management	9 29	7 27	2 11.8	- -	- -	2 11.1	20 17.1
Mathematic tools employed to make models for educational management	3 9.6	2 7.8	1 5.8	- -	- -	- -	6 5.1
IT applications for teaching	- -	1 3.8	2 11.8	- -	3 21.4	2 11.1	8 6.8
Teacher and manager training in the use of IT for educational management	- -	- -	- -	2 18.2	3 21.4	1 5.5	6 5.1
TOTAL	31 100	26 100	17 100	11 100	14 100	18 100	117 100

If, say, we were to consider each different type of application for educational management as a research topic this would probably result in an excessively large group of topics. In this case we have decided to group all such applications in one single category making no distinction regarding the specific applications. We consider that this offers a clearer vision of the attention given to this topic at the different conferences, which would be more difficult to perceive if the analysis were more fragmented. A total of 9 topics were identified, although some have an irregular appearance over time.

The last column of Table 2 reveals that the most popular topics over the years have been Assimilation and integration of IT into educational management (23%), followed by Assessment of IT support to educational management (17.1%), IT applications in educational management (17.1%) and National, regional and local experience in the use of IT for educational management (15.4%). All together these topics represent 76.2% of all papers studied (117). Interestingly, the most recurrent topic over these 10 years of publications has been Assessment of IT support to educational management, which has been repeatedly dealt with because of the importance of analysing the results obtained after implementing new strategies, policies, techniques or tools.

On the other hand, although Assimilation and integration of IT into educational management is the topic which has received most attention (23%) it has progressively lost importance judging by the decreasing number of papers that address it. This is probably due to the fact that a certain maturity has now been reached in ITEM and a solidly built theoretic body of practices and recommendations is now being developed and is obtaining good results.

Also, there are two themes that have followed a diametrically opposed evolution through the years this analysis covers. The topic Mathematic tools employed to make models for educational management of certain importance during the first three conferences, has not been dealt with at all for the last three. This also reveals a strong connection between authors and subjects, so that if certain authors do not take part in the conferences certain topics will not be discussed. Conversely, the theme we call Teacher and manager training in the use of IT for educational management is relatively important in the last three editions whereas it was totally absent from the first three. The reason for this is probably the growing importance of training as a crucial factor for a successful implantation of information technology, a matter that is also present in any IT related field.

A subject that has always been present, albeit discreetly, in the publications that follow the conferences is the one we call Strategies to integrate IT into educational management. This gives evidence of a consensus regarding the need to make an effort to plan future actions, especially if they are far-reaching, independently of technological developments and advancements. Finally, another subject present though not directly related to the field of Working Group 3.7 refers to IT applications for teaching, which has appeared in several publications over the last ten years.

4. RESEARCH METHODS

As regards research methodology, the papers we have reviewed can be divided into theoretical studies and empirical studies. We have grouped the theoretical studies into conceptual and illustrative categories and the empirical ones have been classified as case studies and field studies. We will describe each of these methods briefly so as to understand them better.

Theoretical studies are fundamentally based on ideas, structures and speculations rather than a systematic observation of reality. Although non-empirical articles may contain some empirical observation or fact, these will be of secondary importance. In other words, emphasis is on ideas rather than facts. Theoretical studies can be of a conceptual and illustrative kind. The first describe structures, models or theories and offers explanations and reasons. The illustrative ones, on the other hand, intend to guide practice and make recommendations for action or establish stages in which to attend to certain circumstances. The emphasis is on what and how rather than why.

The essence of research carried out in empirical studies is to observe the reality object of investigation. This is where we can place case studies. These kind of studies are becoming ever more numerous in the field of IS/IT, mostly for the following reasons (Benbasat *et al.*, 1987): (a) the researcher can study IS/IT in their natural environment, learn about the state of the art and generate theories based on practice; (b) it allows researchers to answer questions as to how and why and therefore to understand the nature and complexity of the process that is taking place; and (c) it is appropriate for investigation in areas with few previous studies and is often the first stage of empirical research. But case studies have often been criticised for their lack of scientific rigor, though this is not due to a problem in the method itself but rather to the fact that often the name 'case study' has been given to what is merely a recounting of anecdotes (Lee, 1989).

Field study is another empirical research method that analyses several organisations regarding one or more variables. There is an experimental design, but no experimental control, which means that the researcher collects information concerning uncontrolled situations. The object of study operates in its usual fashion while research is conducted. The aim is to relate results to certain explanatory variables. It is similar to case study in that phenomena is analysed in its natural environment without introducing any variations in it. But the methods differ in that field study is not interested in the whole phenomena but only in specific aspects or variables. Also, the analysis of information in case studies is merely qualitative whereas field studies generally use quantitative methods.

Table 3 shows a classification of papers considering the research methodology employed. Similarly to Table 2, in Table 3 the total number of papers is shown as well as the percentage of each according to the methodology used and progress over time of each method, expressed for each of the publications considered. The last column shows total values.

Table 3: Papers classified by research methodology

Research methodology	CONFERENCE 1994 N. %	1996 N. %	1998 N. %	2000 N. %	2002 N. %	2004 N. %	Total N. %
Theoretical studies	**15** **48.4**	**15** **57.7**	**4** **23.5**	**5** **45.5**	**4** **28.6**	**7** **38.9**	**50** **42.7**
Theoretical – conceptual	6 19.4	4 15.4	2 11.8	1 9	1 7.2	4 22.2	18 15.4
Theoretical – illustrative	9 29	11 42.3	2 11.8	4 36.4	3 21.4	3 16.7	32 27.3
Empirical studies	**16** **51.6**	**11** **42.3**	**13** **76.5**	**6** **54.5**	**10** **71.4**	**11** **61.1**	**67** **57.3**
Empirical – case studies	13 41.9	10 38.5	9 52.9	2 18.2	7 50	8 44.4	49 41.9
Empirical – field studies	3 9.7	1 3.8	4 23.5	4 36.4	3 21.4	3 16.7	18 15.4
TOTAL	31 100	26 100	17 100	11 100	14 100	18 100	117 100

An analysis of Table 3 shows that most of the articles published are empirical, exceeding the theoretical ones by more than ten percentage points. Except for 1996, there is a clear preponderance of empirical studies in all conference publications, which suggests that tendencies regarding research methods in the field of ITEM are quite homogenous. We find a similar situation in the field of IS/IT, in which as from the middle 80s empirical studies prevail over strictly theoretical ones (Alavi and Carlson, 1992).

We can also see that amongst the theoretical studies the illustrative type is the most frequent and very nearly doubles the conceptual kind. This situation is typical in any emerging field in which an ample heuristic is transformed into recommendations and good practice guides based on the author's experiences. Similarly, the fact that most of the empirical studies are case studies shows that we are facing the first stages of research. This supports our previous comment that case studies are especially useful during the explorative phases of any investigation, whereas field study requires a broader understanding of the phenomenon and therefore is better suited to more advanced phases of research (Lai and Mahapatra, 1997).

5. COMPARISON WITH OTHER AREAS OF STUDY

ITEM is a field in which information systems and technology have an outstanding role. This explains the many common characteristics that both fields share, mainly their relative youth compared to other more consolidated disciplines such as Sociology, Psychology, Economics or Computer Science. Both ITEM and IS/IT have benefited from all these fields in some way and particularly ITEM has received important contributions from IS/IT, as with design strategies recommended to develop ITEM applications (Tatnall, 2001).

Therefore it is not surprising that there are so many coincidences between both fields regarding topics of study. We have, for example, the investigation carried out by Claver *et al.* in 1999 concerning the most frequent topics in IS/IT research as regards papers published in *Information & Management* and *MIS Quarterly* between 1981 and 1997. Referring only to the two-year periods 1994/95 and 1996/97, which are the only ones we can compare with ITEM publications for the 1994 and 1996 conferences, there is a great coincidence in the percentage of studies on certain topics, as is shown in Table 4. Such themes are *Assessment* and *Strategies*. There is, on the other hand, only an approximate connection with other topics during part of the given period, such as with *IT Applications*. However, as regards *Assimilation and Integration of IT* results are totally different, probably because this topic had already been sufficiently dealt with in IS/IT literature before the period we have considered.

Table 4: Comparison of scientific production regarding certain common topics in the IS/IT and ITEM fields

IS/IT field			**ITEM field**		
Topic	1994/95	1996/97	Topic	1994	1996
IS assessment	6.5 %	14 %	Assessment of IT support to educational management	6.5 %	19.2 %
IS strategies	5.9 %	0.8 %	Strategies to integrate IT into educational management	6.5 %	3.8 %
IT applications	9.8 %	26.9 %	IT applications in educational management	29 %	27 %
IS alignment with organisational management	3.9 %	10.4 %	Assimilation and integration of IT into educational management	12.9 %	34.6 %

6. CONCLUSIONS

The interest in overcoming problems derived from adapting a new and changing technology such as IT to educational management has set the standards for ITEM research during Working Group 3.7's ten years of existence. In such a brief period of time no less than six books have been published, with the work conferences' main conclusions, that allow us to confirm that this area of study is now reaching its maturity. The topics handled, the employed methods of investigation and the parallelism between ITEM and IS/IT study percentages are all evidence to this.

Nevertheless, ahead of us is the task of completing a compact body of learning made up of theories that can help ITEM acquire its own identity. This is not easy owing to its multidisciplinary nature and its strong dependence on such a changing support such as IT.

This study should be extended by referring to literature on the use of IT applied to educational management other than that published by our specific working group. This would undoubtedly broaden our views on this extremely interesting field of study.

7. REFERENCES

Alavi, M. and Carlson, P. (1992). "A review of MIS research and disciplinary development". *Journal of Management Information Systems*. Vol. 8, N. 4: 45-62.

Benbasat, I.; Goldstein, D.K. and Mead, M. (1987). "The case research strategy in studies of information systems". *MIS Quarterly*. Vol. 11, N. 3: 369-386.

Claver, E.; González, M.R. and Llopis, J. (1999). "Estudio de la investigación en sistemas de información a través del análisis de dos revistas (1981-1997)". *Revista de Economía y Empresa*. Vol. XIII, N. 36: 97-126.

Grover, V.; Lee, C.C. and Duran, D. (1993). "Analizing methodological rigor of MIS survey research from 1980-1989". *Information & Management*. Vol 24, N. 6: 305-317.

Lai, V.S. and Mahapatra, R.K. (1997). "Exploring the research in information technology implementation". *Information & Management*. Vol. 32, N. 4: 187-201.

Lee, A.S. (1989). "A scientific methodology for MIS case studies". *MIS Quarterly*. Vol. 13, N. 1: 33-50.

Tatnall, A. (2001). "Design strategies", in Visscher, A.J.; Wild, P. and Fung, A.C.W. (eds.) *ITEM: synthesis of experience, research and future perspectives on computer-assisted school information systems*. Kluwer Academic Publishers. Dordrecht.

Visscher, A.J.; Wild, P. and Fung, A.C.W. (2001). *ITEM: synthesis of experience, research and future perspectives on computer-assisted school information systems*. Kluwer Academic Publishers. Dordrecht.

Assessment Information Systems for Decision Support in Schools

A Case Study from Hungary

Andreas Breiter & Emese Stauke
Institute for Information Management University of Bremen, Germany

Abstract: Schools as places for institutionalized learning could be a perfectly suitable domain for knowledge management systems. Making knowledge about teaching and learning as well as school performance available to the relevant stakeholders seems to be a promising approach. The crucial question is how to identify information needs, select the relevant data and how to organize feedback. In this paper, the computer-based support of classroom decisions and school management on the basis of standardized test results will be presented. With the help of these assessment information systems, feedback on different levels for different target groups in the school system can be organized. Reflecting on the rich body of empirical research on management information systems, we will present an example for an assessment information system from Hungary, which has just started with computer-support for data-driven decision-making. This case study illustrates the potential added value for the key stakeholders in the school systems.

Keywords: Assessment information system, decision support.

1. INTRODUCTION

During the last decade, knowledge management has also floated from the world of business into the education sector. Regardless of its exaggerated promises, schools as places for institutionalized learning could be a perfectly suitable domain for knowledge management systems. Making knowledge about teaching and learning as well as school performance available to the relevant stakeholders seems to be a promising approach. The crucial question is how to identify information needs, select the relevant data and how to organize feedback. One important sub-group for knowledge management is data-driven decision-making on the basis of data from internal and external evaluations. The shift in educational policy towards accountability has prompted many districts and school administrators to

Please use the following format when citing this chapter:

Breiter, A. and Stauke, E., 2007, in IFIP International Federation for Information Processing, Volume 230, Knowledge Management for Educational Innovation, eds. Tatnall, A., Okamoto, T., Visscher, A., (Boston: Springer), pp. 9–17.

think differently about the potential that assessment data and information systems may have to inform instruction and decision-making aimed at raising student achievement. With international studies like TIMSS or PISA, as well as with national test systems and self-evaluation, data becomes widely available. Increasingly the exploration of how data can inform instructional decisions for school improvement is a major topic of educational policy and building Management Information Systems (MIS) is a central concern for many administrators. The assumption is that these systems can support the decision-making process and play an important role for technological support of knowledge management in schools.

We will focus on information systems to feed back results, what we call "assessment information systems", introducing an example from Hungary. We will draw on organizational research on MIS as well as on studies on school information systems (SIS) to explore the question of how schools use information systems and how the information needs of end users can be mapped across different levels of the system. The case study examines the implementation of an information system for results from the national test in Hungary. Finally, the paper concludes with a discussion about critical factors for the implementation of information systems for schools and the meaning the data can have for data-driven decision-making and knowledge management in schools.

School information systems constitute a sub-group of management information systems that are used in educational organizations. In schools, different information systems support different types of decisions: strategic, tactical and operational (for this distinction see Gorry & Scott Morton, 1971). In principle we must distinguish between systems that are focused directly on the support of the teaching and learning process and systems that serve for managerial, administrative and instructional decisions.

The systematic analysis of assessment information systems for school development will be a determining topic for schools and their management in the next few years. Similarly, Visscher (2002) describes these information systems as school performance feedback systems. In his theoretical framework, they serve two different goals – accountability or school improvement (Visscher, 2002, p. 44).

With this typology in mind, we will focus primarily on assessment information systems for school improvement and its relevance for knowledge management in schools.

2. DATA-DRIVEN DECISION-MAKING AND KNOWLEDGE MANAGEMENT IN SCHOOLS

The terms 'information' and 'knowledge' are often used as though they were interchangeable, when in practice their management requires very different processes. At the core of knowledge management is the idea of systematizing the variety of data and to provide the resulting product in an appropriate form concerning time and place using information technology (e.g. Alavi & Leidner, 2001; Earl, 2001). Thereby it should become possible that users can gain relevant knowledge from internal and external sources. In our case, we will have a closer look at available data in the school context. In most theories of knowledge management, knowledge is regarded as being embedded in people, and its creation occurs in the process of social interaction about information (e.g. Brown & Duguid, 2000): "Information is a flow of messages, while knowledge is created by that very flow of information anchored in the beliefs and commitment of its holder. This [...] emphasizes that knowledge is essentially related to human action" (Nonaka & Takeuchi, 1995). Likewise, Drucker (1989) claims that "[...] knowledge is information that changes something or somebody – either by becoming grounds for actions, or by making an individual (or an institution) capable of different or more effective action" (Drucker, 1989). Therefore, data, prior to becoming information, doesn't speak for itself and is not connected in a meaningful way to a context or situation. Borrowing from Ackoff's (1989) work, we adapted a simplified version of his conceptual framework (see also Breiter & Light, 2004). Within the framework, there are three "phases" of the continuum:

- Data exist in a raw state. They do not have meaning and therefore, can exist in any form, usable or not. Whether or not data become information depends on the understanding of the person looking at the data.
- Information is data that is given meaning when connected to a context. It is data used to comprehend and organize our environment, unveiling an understanding of relations between data and context. Alone, however, it does not carry any implications for future action.
- Knowledge is the collection of information deemed useful, and eventually used to guide action. Knowledge is created through a sequential process. In relation to test information, the teacher's ability to see connections between students' scores on different item-skills analysis and classroom instruction, and then act on them, represents knowledge.

We identified six broad steps through which a person goes in order to transform data into knowledge. The process entails collecting and organizing data, along with summarizing, analysing, and synthesizing information prior to acting. This is the process through which raw data are made meaningful, by being related to the context or situation that produced it; consequently, human action underlies all decision-making. This sequential process underlies our understanding of how teachers interact with data. In the context of the use of assessment data the described process is also depending on the organizational context. According to Weiss (1998), use of evaluation results is relatively simple when the expected changes are closely related to teachers' prevalent practices and congruent to existing organizational processes and stakeholders' values and beliefs (Weiss, 1998). If the expected changes go beyond these expectations, any use of data for school improvement can be problematic.

3. TECHNOLOGICAL SUPPORT FOR DECISION-MAKING IN SCHOOLS

Research about using test data to support classroom-level decisions or building-level planning to improve learning is just beginning to emerge. In the U.S., research from different design sites that are piloting educational data-systems are seen in: the Quality School Portfolio (QSP) developed at CRESST (Mitchell & Lee, 1998), the Texas Education Agency, and the South Carolina Department of Education (Spielvogel & Pasnik, 1999). Research on the role of data systems and applications in practice is being done in Minneapolis (Heistad & Spicuzza, 2003), Boston (Sharkey & Murnane, 2003), and on feedback systems in Louisiana (Teddlie, Kochan, & Taylor, 2002). In the UK and the Netherlands, school performance feedback systems are widely spread, although research on their use is just starting (for an overview see Visscher & Coe, 2002). So far, there is no evidence, that feedback for stakeholders in school make a significant difference. "Feedback can be harmful to performance as often as it improves it. In designing and implementing feedback systems, we should be conscious that they cannot be guaranteed automatically to do good. However, the evidence is also clear that, under the right conditions, feedback can have a substantial effect on improving task performance" (Coe, 2002, p. 22f). Except for the work done by Thorn in Milwaukee (Thorn, 2001, 2003), Visscher in the Netherlands and Coe in the UK (see Visscher & Coe, 2002), there is only little empirical evidence. In a recent study of a web-based assessment information system in New York City (Breiter & Light, 2004; Light et al., 2005), the authors identify four success factors: alignment to information needs of all stakeholders, ownership of data in schools, thoughtful visualization and additional instructional material.

Although research is just beginning in the education field, MIS have been in focus since the early 70s in management research. We can find today in schools a similar situation concerning the implementation and adoption of MIS as in business 20 or more years ago. We can learn from the empirical research in MIS that decision-makers at different levels of the school system have different information needs, which first have to be identified and analysed. In the school system five levels can be distinguished (see table 1).

Level	Stakeholders	Information needs
Classroom	Teachers Students	• Disaggregated student data • Grades and test scores / portfolios • Tracking of attendance / suspensions
School	Principal Administrators	• Aggregated longitudinal student data (i.e. by class, by subject) • Grades and test scores • Tracking of attendance / suspensions • Aggregated longitudinal administrative data • Coordination of class scheduling

Level	Stakeholders	Information needs
		• Special education and special programs scheduling • Allocation of human resources • Professional development • Finance and budgeting
District	Superintendent Administrators	• Aggregated longitudinal student data (i.e. by building, by grade) • Aggregated longitudinal administrative data (i.e. by building, by grade) • External data reporting requirements
School Environment	Parents Local community	• Disaggregated student data • Aggregated administrative data

Table 1: Model of levels of information needs in schools (Breiter & Light 2004)

As extensively discussed in empirical studies of MIS (e.g. Ackoff, 1967; Feldman & March, 1981; Gorry & Scott Morton, 1971; Orlikowski, 1992), key aspects can be extracted that help an understanding of influential factors for decision-support in schools:

(1) Decision-making is a highly complex individual cognitive process influenced by various environmental factors. The classroom may be the example par excellence of an inter-subjective decision-making environment. Teachers constantly make decision affecting 20 or more individuals. (2) Decision-makers often are not fully cognizant of the specific data they rely on for each decision. Identifying the information needs of the decision maker is a crucial step in designing an effective information system, although identifying the appropriate information to put into the system is a complex and time-consuming process. (3) An effective information system needs to incorporate the logistical elements of time, quantity, quality and access. Schools are challenging organizations to work with. There is a lack of administrative staff for data processing and distribution. If the access to data is managed centrally, teachers are often unable to leave the classroom to retrieve information. Hence, the necessary infrastructure in every classroom has to be in place.

4. CASE STUDY FROM IN HUNGARY

4.1 The Hungarian context

Hungary is a central European country and has just joined the EU. According to the Hungarian Central Statistical Office, there are more than 10,000 schools with two million students and 190,000 teachers (HCSO, 2005). The Értékelési Központ (Hungarian Centre for Evaluation Studies, HCES) carries out national and international comparative studies in order to evaluate the school system and to assess student literacy (reading and understanding, mathematics and science). The literacy concept as well as the psychometric model is very closely related to the methodology used in PISA. Since 1986, Hungarian students participate in assessment tests and longitudinal bi-annual studies, which serve for monitoring the performance of a student population. The new test series were started in 2002, directly after the first PISA results for the Hungarian education system were published. The results especially in reading literacy were below average, while the results for mathematics and science were lower than expected. This was the trigger to establish a new quality management system for schools. The tests are carried out for different age groups every year. Their aim is not only the measurement of learning status, but also to find out if students are able to use their school knowledge in authentic contexts. Another aim is the establishment of a local "testing culture" in schools. It is assumed that schools, which gain experience with testing methods and tools and learn how to access the data to make their own assessments, will consequently accept nation-wide testing.

In 2002 the skills of students in grades 5 and 9 were tested in reading, understanding and mathematics at the beginning of the school period. After election in 2003, the system has slightly changed. All Hungarian students in grade 6 (120,000 students) and grade 9 (110,175 students) were tested in reading, understanding and mathematics. As a weighted sample, 20 test booklets from each age group of all schools are sent to HCES. For the first time in history, the competence level of each student was measured and socio-economic variables were included in the evaluation. In 2004, the tests were extended to grades 6, 8 and. The tests took place nationwide at the same day and were complemented by questionnaires on context variables (with a special index for the socio-economic status).

The central sample is coded and analysed centrally by experts. Multiple-choice questions are analysed automatically, while open questions are evaluated in groups in order to reach inter-subjective concurrence. Context data (e.g. socio-economic status and environmental factors in the specific school) are collected with school and student questionnaires. The data of the school questionnaires are not analysed centrally, but provided to schools for self-evaluation. The student questionnaires are assigned to the respective school, and a SES index for the school is developed. With this index, the observed results could be compared with the expected results with regard to the socio-economic background of the student population (Balázs & Zempléni, 2004). Teachers and administrators in each school can analyse the data from the central sample with special software that was sent to schools. They receive their files by using a school-specific code.

4.2 Assessment Information System

The feedback on results is carried out in two different ways. After evaluation, a paper-based individualized report for each school is sent to all participating schools. This report includes details on the following subjects:

- Average results of the school compared to all Hungarian schools, to all urban/rural schools, to cities/villages/schools of similar size
- Distribution of pupil performance in cities/villages/schools of similar size
- Distribution of the points for each item reached by students
- Distribution of results by competence levels
- Result distribution with box plots
- Trend analysis
- Comparison of observed with expected values

All diagrams are explained in detail and substantiated by examples. After having received the software, the schools can either manipulate this data or add manually the remaining test data to have a complete set. The results entered are immediately displayed in various forms of visualization. The entry page provides an overview over the most important information of the evaluation data. Here the user can select between different booklet alternatives, the subject (mathematics or reading) and the grade (6, 8 or 10). The results are presented in three types of diagrams:

1. Average values by bar charts (country average, average by cities/villages, average by cities/villages/schools of similar size)
2. Percentile values compared to rural region, cities/villages or to cities/villages/schools of similar size; the data can be evaluated for the whole test or for individual tasks.
3. Box plots by the presentation of minimum and maximum, the lowest and highest percentile, the lowest and highest quartile and the median.

The ranking of schools or students is not possible. The available data is grouped by test items and not by individuals. During the last testing phase, approximately 20 per cent of schools have used the software for their own evaluation process. An evaluation study on the actual use of the results within the school is currently planned. As schools are the owner of their data, they have the option to evaluate their work without direct external pressure. Additionally, the CD-ROM version of the software provides schools with a “sensed” security, i.e. the data is only available to schools and not via the Internet (or the district).

5. CONCLUSIONS

Drawing on the brief example from Hungary, the potential of computer-based feedback systems to provide assessment data is obvious. Reflecting on other empirical studies, it cannot be taken for granted that use of information systems will lead automatically to school improvement. Organizational change, as well as teacher training and trust building to avoid the unintended consequences have to be taken into account. The Hungarian case shows how thoughtful implementation procedures can help to slowly adapt to the requirements of teachers and administrators. The findings from Hungary offer an illuminating example for other countries that are starting to use test data for accountability and policy making. From a school management perspective, the software can be regarded as a useful data-tool as the school still holds ownership of the data.

There is more research necessary on the design principles of assessment information system, drawing on a large body of research on participatory design to involve all stakeholders an the development process. In addition, there is only little evidence on the actual use of such systems and how they have to be embedded into the organizational context. Answers to these questions can lead to a deeper understanding of knowledge management in schools.

6. REFERENCES

Ackoff, R. L. (1967). Management Misinformation Systems. *Management Science, 14*(4), 147-156.

Alavi, M., & Leidner, D. E. (2001). Knowledge Management and Knowledge Management Systems: Conceptual Foundations and Research issues. *Management Information Systems Quarterly, 25*(1), 107-136.

Balázs, I., & Zempléni, A. (2004). A hozottérték-index és a hozzáadott pedagógiai érték számítása a 2003-as kompetenciamérésben. *Új Pedagógiai Szemle, December*, 36-50.

Breiter, A., & Light, D. (2004). *Decision Support Systems in Schools – from Data Collection to Decision Making.* Paper presented at the America's Conference on Information Systems (AMCIS), New York, NY.

Brown, J. S., & Duguid, P. (2000). *The Social Life of Information.* Boston, MA: Harvard Business School Press.

Coe, R. (2002). Evidence on the Role and Impact of Performance Feedback in Schools. In A. J. Visscher & R. Coe (Eds.), *School Improvement Through Performance Feedback* (pp. 3-26). Lisse: Swets & Zeitlinger.

Drucker, P. F. (1989). *The New Realities: In Government and Politics. In Economics and Business. In Society and World View.* New York, NY: Harper & Row.

Earl, M. J. (2001). Knowledge Management Strategies: Towards a Taxonomy. *Journal of Management Information Systems, 18*(1), 215-233.

Feldman, M. S., & March, J. G. (1981). Information in Organizations as Signal and Symbol. *Administrative Science Quarterly, 26*, 171-186.

Gorry, G. A., & Scott Morton, M. S. (1971). A Framework for Management Information Systems. *Sloan Management Review, 13*(1), 55-70.

HCSO. (2005). *Data of Education (Preliminary data).* Retrieved Sep 30th, 2005, from http://portal.ksh.hu/portal/page?_pageid=38,248221&_dad=portal&_schema=PORTAL

Heistad, D., & Spicuzza, R. (2003, April). *Beyond Zip code analyses: What good measurement has to offer and how it can enhance the instructional delivery to all students.* Paper presented at the AERA Conference, Chicago.

Light, D., Honey, M., Heinze, J., Brunner, C., Wexlar, D., Mandinach, E., et al. (2005). *Linking Data and Learning - The Grow Network Study. Summary Report.* New York: EDC's Center for Children and Technology.

Light, D., Wexler, D., & Heinze, J. (2004). *How Practitioners Interpret and Link Data to Instruction: Research Findings on New York City Schools' Implementation of the Grow Network.* Paper presented at the Conference of the American Educational Research Association (AERA), San Diego, CA.

Mitchell, D., & Lee, J. (1998). *Quality school portfolio: Reporting on school goals and student achievement.* Paper presented at the CRESST Conference, Los Angeles, CA.

Nonaka, I., & Takeuchi, H. (1995). *The Knowledge Creating Company*. Oxford: Oxford University Press.

Orlikowski, W. J. (1992). The duality of technology: Rethinking the concept of technology in organizations. *Organization Science, 3*(3), 398-427.

Sharkey, N., & Murnane, R. (2003). *Helping K-12 Educators Learn from Student Assessment Data*. Paper presented at the AERA, Chicago.

Spielvogel, B., & Pasnik, S. (1999). *From the School Room to the State House: Data Warehouse Solutions for Informed Decision-Making in Education*. New York: EDC/Center for Children and Technology.

Teddlie, C., Kochan, S., & Taylor, D. (2002). The ABC+Model for School Diagnosis, Feedback, and Improvement. In A. J. Visscher & R. Coe (Eds.), *School Improvement Through Performance Feedback* (pp. 75-114). Lisse: Swets & Zeitlinger.

Thorn, C. A. (2001). Knowledge Management for Educational Information Systems: What is the State in the Field? *Education Policy Analysis Archives, 9*(47).

Thorn, C. A. (2003). Making Decision Support Systems Useful in the Classroom: Designing a Needs Assessment Process. In I. Selwood, A. C. W. Fung & T. Paturi (Eds.), *Information Technology in Educational Management*. Dordrecht: Kluwer Academic Publishers.

Visscher, A. J. (2002). A Framework for Studying School Performance Feedback Systems. In A. J. Visscher & R. Coe (Eds.), *School Improvement Through Performance Feedback* (pp. 41-72). Lisse: Swets & Zeitlinger.

Visscher, A. J., & Coe, R. (Eds.). (2002). *School Improvement Through Performance Feedback*. Lisse: Swets & Zeitlinger.

Weiss, C. H. (1998). Improving the use of evaluations: whose job is it anyway? In A. J. Reynolds & H. J. Walberg (Eds.), *Advances in educational productivity, Volume 7* (pp. 263-276). Greenwich: JAI Press.

Research Knowledge Management can be Murder

University Research Management Systems

Bill Davey and Arthur Tatnall

School of Business Information Technology, RMIT University, and Centre for International Corporate Governance Research, Victoria University, Australia

Abstract: Use of the term *knowledge management* varies depending on the context. In this paper we will investigate its use in relation to research output in universities. Universities need to keep track of their research and to note what research papers have been written, the topic of the research, who collaborated in the writing and where the research was published. To collate and store this data some type of information system is needed. There are several reasons why universities need to keep track of their research output. Firstly this is necessary for accountability purposes, and in order to gain funding from Governments and other funding bodies – this is the principal reason why such systems are set up. Universities also like to publicise what they are doing and this also requires recording research output. Another possible use of this information however, relates to intellectual capital and the management of knowledge. Researchers can benefit greatly from knowing what other researchers have done, and what they are currently doing, but universities are large institutions making this difficult to achieve. In the paper we will argue that this constitutes an important but underutilised application for research management systems.

Keywords: Research, information management systems, accountability, quality, knowledge management.

1. INTRODUCTION

In 1992 a Canadian academic called Valery Fabrikant, went into work carrying three handguns and a briefcase full of ammunition, and killed four of his colleagues. Amongst other issues reported about the case were his allegations that the university tolerated widespread academic fraud (Spurgeon 1994a). A subsequent investigation found that research irregularities had indeed taken place in the university and had involved several academics including the murder victims (Rolston 1994). The investigators reluctantly noted that Valery Fabrikant was correct in his

Please use the following format when citing this chapter:

Davey, B. and Tatnall, A., 2007, in IFIP International Federation for Information Processing, Volume 230, Knowledge Management for Educational Innovation, eds. Tatnall, A., Okamoto, T., Visscher, A., (Boston: Springer), pp. 19–25.

accusations (if not in his actions), noting that he had collaborated with three other Engineering academics in the submission of the same academic paper to several different academic journals in the USA, Germany, France, and Britain. The report noted that all the papers were "quite extraordinarily similar" to work that Fabrikant had originally published in 1971 in an obscure Russian journal. For what amounted to plagiarism, it blamed an over-competitive research atmosphere in which academics were valued by how often they published.

One would hope that murder is not a prerequisite for universities to examine their research quality and the data gathered to determine research performance. This paper seeks to set minimum standards for a knowledge management system required for research quality in an academic institution. We argue that to use these systems *only* for accountability and funding purposes means missing out on a much of their potential value (Tatnall and Tatnall 2006 forthcoming).

2. THE NEED FOR MANAGEMENT OF RESEARCH KNOWLEDGE

There is not a single agreed meaning of the term 'knowledge management' (Cader 2004). We will use the term to mean both: "The explicit control and management of knowledge within an organisation aimed at achieving the company's objectives" (Van der Speck and Spijkevert 1997) and "Formalisation and access to experience, knowledge and expertise that create capabilities, enable superior performance, encourage innovation and enhance customer value" (Beckman 1997).

A university plays many roles in the community. These include education of professionals, education required to perpetuate and advance knowledge in a range of disciplines and performing research so as to create new knowledge. Although business models are used by university administrators to help administer the money flows in the organisation a university is not fundamentally based on the assets of buildings and student fee structures. The university has its place in society because each individual academic produces intellectual property that is expressed in courses that are rich in content or research that expands human knowledge. More than in any other sector, a university has only one important capital and that is the knowledge held and new knowledge created by its academic staff.

3. PRESSURES FOR RESEARCH MANAGEMENT

As with many information management systems in education, research information systems often stem from the requirements of funding agencies demanding accountability (Sessions and Collins 1988; Spurgeon 1994b; Tatnall 1995), and in the forward to a discussion paper on research an

Australian Government Minister comments that research should be seen as a key element of an innovative and economically prosperous nation (DEST 2005).

The work reported here involved analysis of research knowledge management systems in New Zealand, UK, Australia and Hong Kong. These systems were chosen because of their common university cultural heritage, common reliance on centrally funded research and similarities in administrative structures. Keeping these environmental factors constant enabled the analysis to concentrate on the relationship between knowledge management systems and research, and the outcomes for universities. Analysis was conducted on the documentation available from several universities in each system and the documentation provided publicly by the funding bodies (in most cases the Central Government). The document analysis was supported by ad hoc interviews with academic and administrative staff in several universities.

3.1 United Kingdom

Under the British Research Assessment Exercise (RAE), research output is considered to be: "any form of publicly available assessable output embodying the outcome of research, as defined for the RAE" (Roberts 2003). The RAE funds traditional universities heavily through a competitive system supported by peer review.

Roberts, who was the principal author of the latest review of university research in the UK in 2001 describes the system as having "evolved from a quality assurance process to a competition for funding, while successfully retaining its original function of driving up standards through reputational incentives". He notes that it has also enabled funds to be concentrated in academic departments best able to produce high quality research.

3.2 New Zealand

The system in New Zealand, mostly created by work done in 2002, is called the Performance-Based Research Fund (PBRF). This funding scheme is based around peer review, both internal and external, and focuses on work done by clusters of researchers. The Minister of Education (representing the funding body) describes the funding scheme as follows: "Focused specialisation, collaboration and co-operation are essential features of a thriving and successful culture of research excellence. The PBRF allows us to reward research excellence and move away from a crude 'bums on seats' approach to funding research which was based on student numbers." The scheme is focused, meaning that there are research winners, and that new researchers must become part of a focus group (New Zealand Ministry of Education 2002).

3.3 Australia

A Commonwealth Government research funding quality scheme is currently being developed for implementation in 2008. This scheme will be used to fund further research in Australian universities. It will include peer review, be based on research centres in universities (or across several universities) and be driven by performance indicators. The scheme allows for separate funding for infrastructure and research training. Research Plans and Research Training Management Plans must be produced and will be externally reviewed (Australian Vice Chancellors' Committee 2005).

3.4 Hong Kong

Since 1993, the Hong Kong University Grants Committee (UGC) has adopted a model that relates the level of funding allocation to the tasks that each institution is expected to accomplish during the funding period, and also to the quality of its recent performance. This determination then provides part of the recurrent funding of each university. The RAE is based on peer review by panels and the effect of the RAE process by the UGC is to concentrate funding less than in the other systems studied. The UGC emphasises that the research assessment exercise "does not imply an interest in research to the possible detriment of teaching quality", and goes on to point out that both teaching and research are important inter-related elements in higher education (Hong Kong University Grants Commission 2005).

4. PROBLEMS WITH CENTRALISED ASSESSMENT SYSTEMS AS A BASIS FOR KNOWLEDGE MANAGEMENT

A common theme that was found in the countries surveyed was the gap between local gathering of data and central use of that data (Tatnall and Pitman 2002). The murder of academics introducing this paper led to an investigation that showed a total lack of *local* use of the research knowledge management information that had been collected. In the years since the murder our investigations have found very little change in this situation despite dire warnings from the research quality literature and the knowledge management literature (Sandy and Davey 2005).

Martin (2000a) notes that a major source of contention in knowledge management is the limited ability of conventional accounting techniques to cope with such intangibles as research and development and with employee talent. Problems associated with poor knowledge management often manifest themselves in outcomes such as a loss of organisational knowledge, expensive duplication of knowledge-creation and acquisition activities, rising costs and reduced competitiveness (Martin 2000b).

The Roberts report contains a salutary warning: "More important, I urge the funding councils to remember that all evaluation mechanisms distort the processes they purport to evaluate" (Roberts 2003).

5. PROBLEMS WITH RESEARCH CENTRES OF EXCELLENCE

One vision of worthwhile research management imagines a group of researchers of sufficient size and resources that enable significant progress to be made in a specific area. In the UK, Australia, and New Zealand the funding bodies have decided to specifically reward individual institutions that create a small number of research units with concentrated resources.

This view imagines a group of researchers of sufficient size that some major research project can be mounted and that peer support in a localised area will lead to research 'greater than the sum of its parts'. This argument is probably justified, but ignores all the other outcomes necessary to create the knowledge required to create a university. Pressing examples of research that can be overlooked by the 'large unit' model include:

- Research by individuals that contribute to highly specialised subjects being taught.
- Research by groups very widely spread and which must span a number of funding areas to provide enough academics to reach critical mass. An example of this can be found in the various IFIP working groups.
- Research from a new area that has not been performed before, and so does not attract experienced researchers. Most topics have had a period in their history where only a few researchers were concerned with the topic.

All of the cases underlying the research in this paper have used peer review as the basis of measurement of research quality. This is exemplified by the RAE in the UK which is essentially a peer review process (Roberts 2003). Similarly in the United States the Agricultural Research, Extension, and Education Reform Act of 1998 (Public Law 105-185) developed a national peer review framework in which all ARS research will be reviewed every 5 years (Knipling 2002).

6. FEATURES REQUIRED IN A RESEARCH KNOWLEDGE MANAGEMENT SYSTEM

Commentators coming primarily from the viewpoint of satisfying funding bodies have a narrow view of what might constitute the requirements of a research information system. In much the same way that some educational administrators see student data purely in terms of school administration and so miss out on the other uses to which such data could be put (Newton and Visscher 2003; Tatnall and Tatnall 2006), these university administrators also ignore the knowledge management advantages inherent in these systems. Like other educational management information systems this has a good deal to do with who the systems were primarily designed for (Tatnall and Davey 2001). To information systems analysts and designers the client lays down the rules on what is required, and if the client is seen as the *administrator* and not the *teacher* then it is not to be expected that the teachers' needs will be considered.

We suggest that in addition to features that enable accountability and funding, an important feature that could be added to a Research Knowledge Management System would facilitate putting researchers in touch with others working on topics that are possibly related to their own. For instance, the authors of this chapter are both academics working from Faculties of Business, but with an interest in education. As they are at different campuses we have very little to do with academics in Schools of Education in our universities and so would not know if they were working in similar areas. It sounds such a small thing, but a system that would make it easy for us to find out what other related work was being done in our own institutions would be very valuable.

A research management system could achieve this in several different ways. For example it could require that each piece of recorded research be accompanied by a number of pre-selected keywords from a list devised by the university and intended to cover a range of research areas. It could then produce custom reports on all the work relating to specified keywords and distribute these to the academics concerned. Not all 'hits' would be relevant, but overall such a system could be very useful.

Similarly, it is easy to overlook other aspects of research not concerned with accountability. Roberts, for instance, identifies: "the need to fully recognise all aspects of excellence in research (such as pure intellectual quality, value added to professional practice, applicability, and impact within and beyond the research community)" (Roberts 2003).

7. CONCLUSION

While we cannot ignore the need to be accountable to funding bodies for monies spent on research, knowledge management within an educational institution has additional needs. To properly manage the growth of human capital a Knowledge Management System must inform the manager of the increase in research output, the emerging new research areas and be able to add research value.

An obvious possibility missing from many such systems is an added capacity to allow researchers within an educational institution to find others with similar areas of research interest. Another is to allow flexible collaborations between universities that are geographically dispersed, as with those in IFIP Working Group 3.7.

8. REFERENCES

Australian Vice Chancellors' Committee (2005). The Research Quality Framework: the AVCC Proposal Developed. Canberra, AVCC.

Beckman, T. (1997). A Methodology for Knowledge Management. AI and Soft Computing Conference, Banff, Canada, International Association of Science and Technology for Development (IASTD).

Cader, Y., Ed. (2004). Knowledge Management: Theory and Application in a Twenty-First Century Context. Melbourne, Heidelberg Press.

DEST (2005). Research Quality Framework: Assessing the Quality and Impact of Research in Australia. Issues Paper. Canberra, Department of Education Science and Training.

Hong Kong University Grants Commission (2005). Research Assessment Exercise 2006. Hong Kong, University Grants Commission (Hong Kong).

Knipling, E. B. (2002). "Forum: National Peer Review Process Sharpens our Science." Agricultural Research. 50(5): 2.

Martin, W. J. (2000a). "Approaches to the Measurement of the Impact of Knowledge Management Programmes." Journal of Information Science 26(1): 21–27.

Martin, W. J. (2000b). "Knowledge Based Organizations: Emerging Trends in

Local Government in Australia." Journal of Knowledge Management Practice(Oct): 21–27.

New Zealand Ministry of Education (2002). Building a Peer Review System for Assessing Research Performance. Wellington, Performance Based Research Fund - Working Group.

Newton, L. and Visscher, A. J. (2003). Management Systems in the Classroom. Management of Education in the Information Age: The Role of ICT. Selwood, I., Fung A. C. W. and O'Mahony C. D. Massachusetts, Kluwer Academic Publishers / IFIP: 189-194.

Roberts, G. (2003). Review of Research Assessment. London, Higher Education Funding Council of England.

Rolston, B. (1994). Concordia Admin Bungled Fabrikant Affair. The Peak - Simon Fraser University's Student Newspaper. Burnaby, British Columbia, Canada: 1.

Sandy, G. and Davey, B. (2005). Data Quality in Educational Systems for Decision Makers. Information Technology and Educational Management in the Knowledge Society. Tatnall, A., Visscher A. J. and Osorio J. New York, Springer: 111-120.

Sessions, R. and Collins, T. (1988). "More Accountability in Federally Funded Academic research: a Costly "Bill of Goods"." Society of Research Administrators 20(1): 195.

Spurgeon, D. (1994a). "Audit backs jailed professor's allegations." Nature 370(6486): 166.

Spurgeon, D. (1994b). "University Censured over Research Accounting." Nature 370(6488).

Tatnall, A. (1995). Information Technology and the Management of Victorian Schools - Providing Flexibility or Enabling Better Central Control? Information Technology in Educational Management. Barta, B. Z., Telem M. and Gev Y. London, Chapman & Hall: 99-108.

Tatnall, A. and Davey, B. (2001). Open ITEM Systems are Good ITEM Systems. Institutional Improvement through Information Technology in Educational Management. Nolan, P. Dordrecht, The Netherlands, Kluwer Academic Publishers: 59-69.

Tatnall, A. and Pitman, A. (2002). Issues of Decentralization and Central Control in Educational Management: the Enabling and Shaping Role of Information Technology. TelE-Learning: The Challenge of the Third Millennium. Passey, D. and Kendall M. Assinippi Park, Ma, Kluwer Academic Publishers / IFIP: 233-240.

Tatnall, C. and Tatnall, A. (2006). Using Educational Management Systems to Enhance Teaching and Learning in the Classroom: an Investigative Study. Elsewhere in this publication.

Van der Speck, R. and Spijkevert, A. (1997). Knowledge Management: Dealing Intelligently with Knowledge. Knowledge management and its integrating elements. Liebowitz and Wilcox. USA, CRC Press.

A Supporting System of Informatics Education for University Freshmen

Manabu Sugiura and Hajime Ohiwa
1 Graduate Schools of Media and Governance, Keio University
2 Faculty of Environmental Information, Keio University

Abstract: Informatics Education for high schools started in 2003 in Japan. Although it is compulsory, university freshmen in 2006 are expected to have large differences in their computer literacy competency. One of the reasons for this is that many teachers have no technical knowledge of Informatics because they were licensed by taking only ninety hours of training during their summer vacation. It is necessary for universities in Japan to classify their freshmen with respect to their computer literacy to be able to provide appropriate education. We have developed a testing system for this but further improvement is required due to quality variations of the test. Appropriate classification of the problems is also required for giving good advice for the student to learn properly. We are developing a Learning Management System (LMS), which supports not only students but also the collaborative works of teachers as well.

Keywords: Informatics Education, Testing System, Learning Management System (LMS), Collaborative Works

1. INTRODUCTION

In 2003, Informatics Education for high schools started in Japan. Although it is compulsory, university freshmen in 2006 are expected to have large differences in their computer literacy competency. One of the reasons for this is that many teachers have no technical knowledge of Informatics because they were licensed by taking only ninety hours of training in during their summer vacation.

These changes have prompted the Keio University Shonan Fujisawa Campus to modify the curriculum related to information processing in 2004. The class of basic information processing was abolished and a qualification test was introduced to measure the competency level of computer literacy.

Please use the following format when citing this chapter:

Sugiura, M. and Ohiwa, H., 2007, in IFIP International Federation for Information Processing, Volume 230, Knowledge Management for Educational Innovation, eds. Tatnall, A., Okamoto, T., Visscher, A., (Boston: Springer), pp. 27–32.

The qualification test consists of three parts: touch typing, basic operation of office applications, and a multiple-choice test for measuring basic knowledge of Informatics.

As a result, teachers had to create many questions for the qualification test and provide a self-learning environment for students. However, there is still an insufficient number of questions and students cannot efficiently study for the qualification test due to the lack of a good self-learning environment. To solve these problems, we propose a Learning Management System (LMS) that not only offers a self-learning environment to students but also supports the managerial work of teachers.

2. CONCEPT OF THE SUPPORTING SYSTEM

The supporting system that we propose supports the following two processes:

1. Process of Self-learning (for the Students)
2. Process of teaching materials management through collaborative works (for the Teachers)

We plan to develop this system by adding new features to the existing LMS. The system is divided into two modules. A relationship of the two modules and details of support are given in Figure 1.

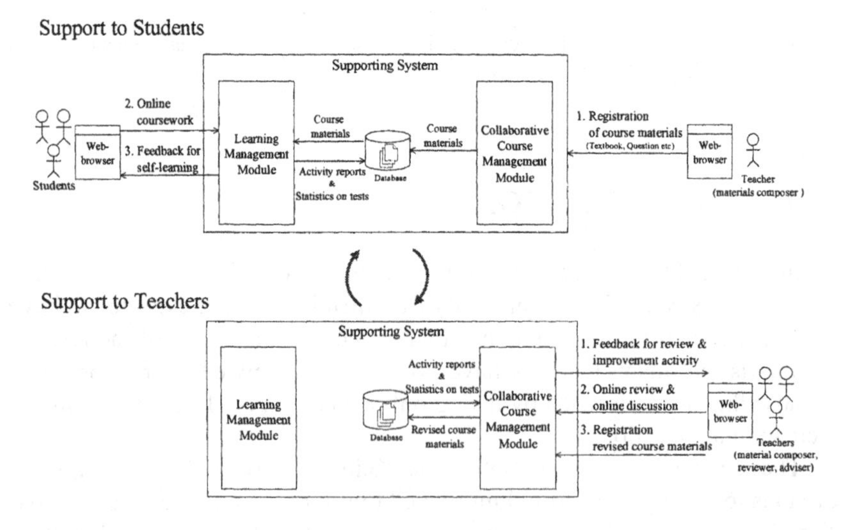

Figure 1. Basic concept of the supporting system

A Learning Support Module supports activities that students learn by themselves using an online textbook and online test. The results of the tests and activity reports of self-learning are stored in the database.

A Collaborative Course Management Module supports teacher activities such as online reviews and online discussion about teaching materials.

2.1 Self-learning process support

By using the existing LMS, teachers can deliver the teaching materials online to students, so we will not need to develop new systems in the future. However, it is necessary to define the self-learning process for the freshmen. Students study according to this process (Figure 2).

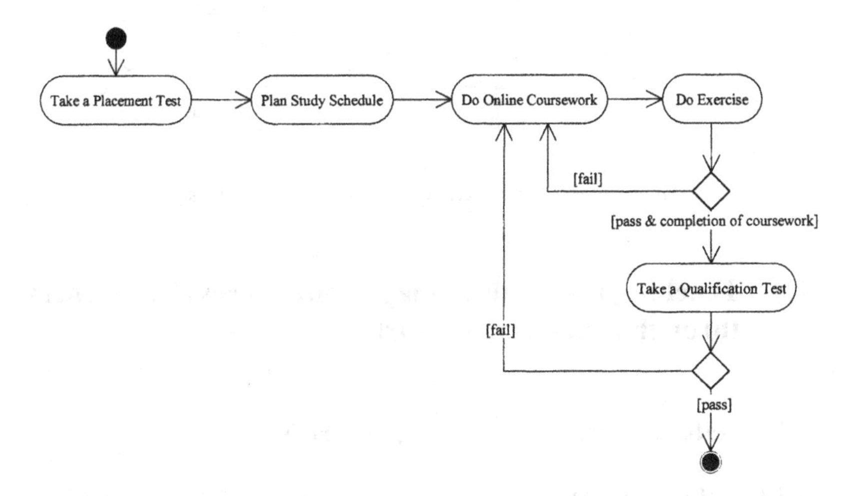

Figure 2. Activities of self-learning

2.1.1 Online placement tests

The mechanism to measure the level of computer literacy is necessary because the freshman should make his/her own study-plan. We think the online placement test is effective as the measurement mechanism. The student plans his/her schedule of self-learning based on the result of the online placement test.

2.1.2 Teaching materials for self-learning and feedback from the system

According to PSI (Personalized System of Instruction) advocated by Keller, students study at their own pace, and do the exercise (Keller 1968). PSI is good match with self-learning that uses LMS (Watanabe & Furukawa 2005). The freshmen are expected to have large differences in their level of computer literacy competency, so they should study at their own pace.

The qualification test result can include review materials if the system has the mechanism for feedback to be included in the qualification test results. A student can be provided with an effective review by showing a link to the textbook relating to the incorrectly answered question (Figure 3).

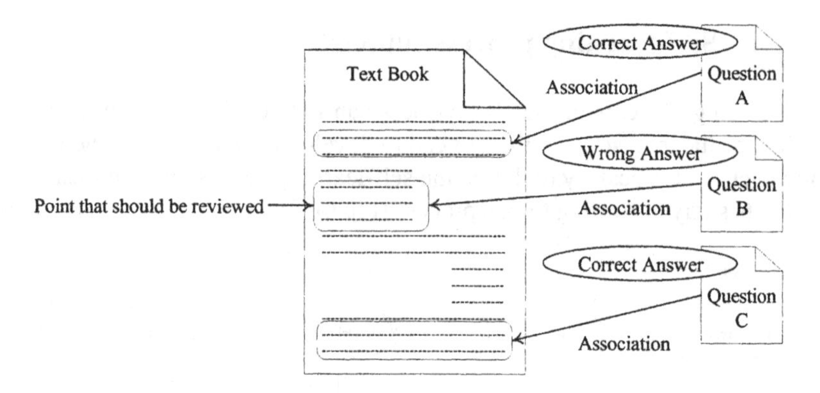

Figure 3. Association between Textbook and Questions

2.2 Teaching materials management process by teachers through collaborative works

2.2.1 State management of teaching materials

Existing Learning Management Systems support uploading teaching materials to the internet. We think that our supporting system should also support the development of teaching materials. The questions on the qualification test and other teaching materials for freshmen should be revised and improved as necessary based on a review by teachers. Now teachers use E-mail as a communication tool; however, this method makes state management of questions difficult.

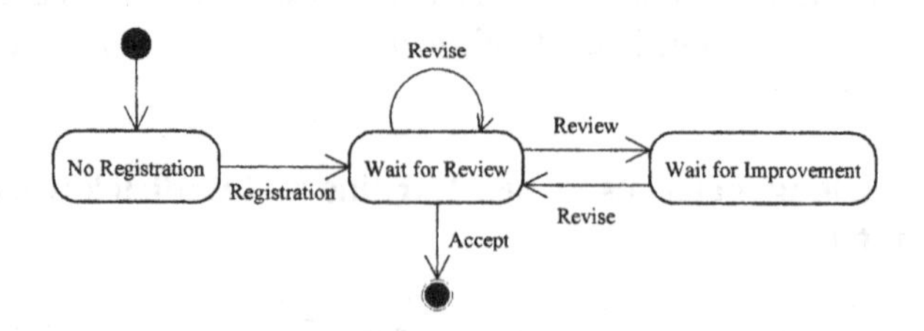

Figure 4. State transition of a question

Figure 4 illustrates a typical state transition of a question. The workflow of developing teaching materials such as questions can be represented by this state transition diagram. The supporting system can define the state of each type of teaching material, transitions and actions. The Content Management Framework (CMF) for Zope provides a tailorable platform for building content management applications (Zope 2005). We think that a framework similar to CMF is required for our supporting system.

Table 1. The features of each tool

	E-mail	BBS	Wiki (2005)	Nongnu (2005)
State management	C	C	C	C
Revision history management	C	C	B	A
Easy to communication (review & discussion)	A	A	A	C
Access control by user authority	C	C	C	B

A: Good, B: Fair, C: Bad

Table 1 gives the features of each tool that can manage teaching materials online. The table shows that E-mail, bulletin board system (BBS) and Wiki are useful for discussion, but on the other hand it is difficult to manage states of materials.

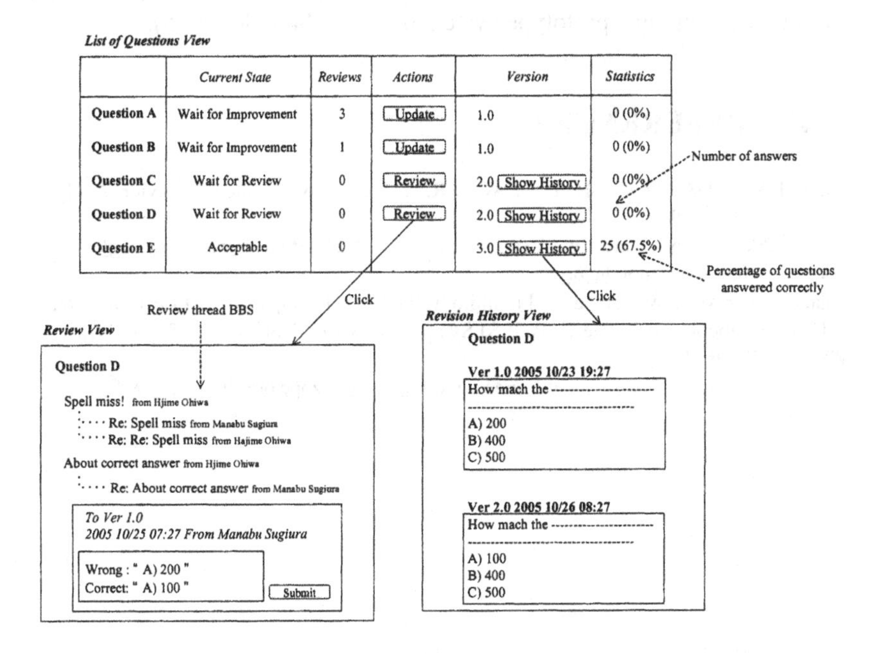

Figure 5. Screen images of question management

Figure 5 is screen images of the management of questions. The supporting system that we propose can manage development of questions by using a Web-browser. The Review view has a BBS because it is suitable for discussion and review. If a question is being reviewed by someone, the current state of the question changes to "wait for improvement" automatically.

2.2.2 Centralization of information management

The student's study records and the results of tests must be stored in our supporting system. This information is used for reference for teaching materials improvement and must be stored in the same database as that of the supporting teaching materials management process.

3. CONCLUSION AND FUTURE WORKS

The system described in this paper offers a self-learning environment to students but also supports the collaborative works of teachers. Using the system, online teaching materials can be efficiently managed by teachers while offering an effective self-learning environment for students.

We wish to analyze the teaching materials being used and define the relationship between them. Creating an accurate placement test is another necessary component to improve the current system. In the search for a flexible system that reduces the managerial works of teachers, we are currently developing a prototype system based on Moodle (2005).

4. REFERENCES

Keller, F.S. (1968). GOOD-BYE TEACHER.... Journal of Applied Behavior Analysis, Vol.1, No1, pp.79-89.

Nongnu (2005). CVS - Concurrent Versions System. (2005). http://www.nongnu.org/cvs/

Moodle. (2005). http://moodle.org/

Watanabe, Hiroyoshi & Furukawa, Fumihito. (2005). An Examples of Self-learning Course Using Textbook and Online Tests. IPSJ Symposium Series, Vol.2005, No8, pp.93-98.

Wiki. (2005). http://wiki.org/

Zope (2005). Content Management Framework. http://www.zope.org/Products/CMF/

LAPCHAT: A Contents-Sharable Management System for Computer Supported Collaborative Learning

Mizue Kayama and Toshio Okamoto
Senshu University, School of Network and Information
Graduate School of Information System, The University of Electro-Communication

Abstract: The purpose of this study is to explore the architecture for a collaborative learning environment, in which individual learning and collaborative learning are smoothly connected. We proposed a composition model of a collaborative workplace, and a management model for action in the collaborative learning space and the state of learning context. Based on the ides, in this paper we describe a Contents-Sharable mechanism between a private workplace and a collaborative workplace.

Keywords: Collaborative learning, learning contents sharing, contents-sharable.

1. INTRODUCTION

In this study, we explore the design of a collaborative learning environment which is realized by the seamless linkage of individual learning and the collaborative learning. A composition model of the collaborative workplace and a management model for learning activity and learning condition have been proposed.

In this paper, the concept of "Contents-Sharable" and a learning system for managing Contents-Sharable are proposed. The sophisticated mechanism for contents sharing, which realize a smooth linkage between learning contents in the private/collaborative workplace, is described. To support these linkages a leaning tool, the "Lapchat", which supports learning in the ICT environment, is described. In this tool, the user controls various kind of information by unification of a type of image. Preparation of the electronic notebook in the ICT environment for individual learning/collaborative or learning/lecture is realized in an OS independent application.

Please use the following format when citing this chapter:

Kayama, M. and Okamoto, T., 2007, in IFIP International Federation for Information Processing, Volume 230, Knowledge Management for Educational Innovation, eds. Tatnall, A., Okamoto, T., Visscher, A., (Boston: Springer), pp. 33–42.

2. THE ARCHITECTURE MODEL OF THE COLLABORATIVE LEARNING FRAMEWORK

In this paper a classification for the architecture of a collaborative learning framework is studied. The MVC (Model–View–Controller) pattern is utilized as a method of the classification. In an attempt at classification, comparisons from the following viewpoints are tried.

- A realization of the synchronous / asynchronous collaborative learning workplace (a private workplace and a collaborative workplace).
- The convenience of adaptation to the educational situation.
- The network traffic load.
- The simplicity of the packaging.

The collaborative learning framework is classified as 4 types based on the management form of the MVC pattern.

2.1.1 Centralized Architecture

The first type is a structure of centralized architecture. In the Centralized Architecture type, only one original application is run on the specific client terminal. All original part (Model, View and Controller) will exist in that specific terminal. The operation event for the original application is transmitted to other terminals. Then, the original GUI (= the View and the Controller) is copied to the specific terminal, and delivered to other terminals. The frameworks that have adopted this architecture are synchronous collaborative learning support systems. Many of these systems have a desktop sharing function. The main feature of this architecture is to work the terminals which run the same OS. The screen image and the operation for any application can be shared between these terminals. However, broadband networks are required in the utilization. The system cannot work when there is a network failure in order to control the whole management by only one terminal.

2.1.2 Replicated Architecture

The second type is a structure of replicated architecture. In this type, all modules related to an application are installed in each client terminal. The original MVC exists in all terminals. Application of this architecture is implemented using the API (Application Programming Interface) for collaboratively utilizing the application. By utilizing the peculiar API, the synchronous function between applications on the each client is secured. The examples of this type are Habanero (Jackson 1999), MatchMaker (Zhao & Hoppe 1994), SAILE (Goodman, Geier, Haverty, Linton & McCready 2001), SimPLE (Plaisant, Rose, Rubloff, Salter & Shneiderman 1999), MediaFusion. In the Replicated Architecture type, the necessary network bandwidth decreases. In addition, the client software can be use as a stand-alone application.

2.1.3 Distributed Architecture

The third type is a structure of distributed architecture In the Distributed Architecture type, MVC is dispersed to multiple hosts. A Model is put to a server. Each View and Controller is allocated to each client: the View/Controller of the client will manipulate the Model on the server. A concrete system of this type is a web site with a database, for example CSILE/Knowledge Forum (Bereiter 1997). The web browser works as the View and the Controller. The server with Model manages data from the Controller and the data for the View. In an application of this type, the View and the Controller are actually constructed on the server. Then, they are transmitted to the client in the execution. This type has common features with the Centralized Architecture type. Additionally, as well as the Replicated Architecture type, the necessary bandwidth is also narrow. The consistency of the Model is also easy to keep. However, the system cannot be utilized as well as the Centralized Architecture type, when there is a failure in the network.

2.1.4 Hybrid Architecture

The fourth type is hybrid architecture. This type is combination of the other forms. For example, a form that combined the Replicated Architecture and the Distributed Architecture was considered. In this type, the synchronization between clients is realized by utilizing the eternal Model on the server (Constantino-gonzalez, Suthers & Santos 2003; Kayama & Okamoto 2002). The application is formed by the MVC on the each user terminal. The condition of the application is preserved in the local file system, when this system works by using each Model as stand-alone. The external Model on the server is utilized for the case of collaborative learning activity. The renewal of operation result of each GUI is guaranteed between clients who login to the server. This type can be utilized even in the situation of failure in the network. However, in this type, the multiple clients are able to renew the common Model. It is necessary to consider data loss consistency in the application implementation.

3. CONTENTS-SHARABLE MANAGEMENT SYSTEM FOR COLLABORATIVE LEARNING

We have explored the design and implementation of a collaborative learning framework. As a result, a learning framework which supports synchronous/ asynchronous collaborative learning using a Hybrid Architecture was proposed. Features of this framework were the integrated mechanism for learning resources sharing, and an event data management mechanism for the record/replay/reference of the collaborative learning situation (Kayama & Okamoto 2002). In order to develop learning resources for this framework, the engineer was required to change his software code by using the specific API. In this framework, the operation history and maintenance of the application's internal state (= the learning condition) were guaranteed by the learning management server. Applications which were rebuilt by using the specific API become the learning resources for our framework. This work is an obstacle for development.

Then a more sophisticated collaborative learning tool was proposed. The features of this tool are to guarantee natural individual learning and to realize the easy linkage and easy management of the group/individual learning result. In this tool, the sharing of learning contents in the private workplace and the collaborative workplace is realized in the form of a contents "image". In this study, this function is named "Contents Sharable".

3.1 The Structure of the Contents-Sharable System

The proposed system is implemented as a Hybrid Architecture that combined the Replicated Architecture and the Distributed Architecture. The outline is shown in figure 1. To each learner terminal, each MVC element is offered. A Model as a eternal object is implemented on the server. The operation by the learner terminal (the Controller data) broadcasts to other learner's terminals via the server. Transmitted information is a set of the Controller data. The Model data is not transmitted. Therefore, in this type, the network load does not rise in comparison with the other Hybrid Architecture. The function of the Contents-Sharable does not depend on the platform of the private workplace and the collaborative workplace. The Model on each learner terminal individually reproduces the condition of the learning situation through the JNI (Java Native Interface) layer. This function works in any user terminal platform, which is able to run the JVM (Java Virtual Machine).

The object structure of the Lapchat is as follows. The client communicates with the server through the connector object. These communications have been realized by distributed object programming. The server involves the Web server. The communication between clients and

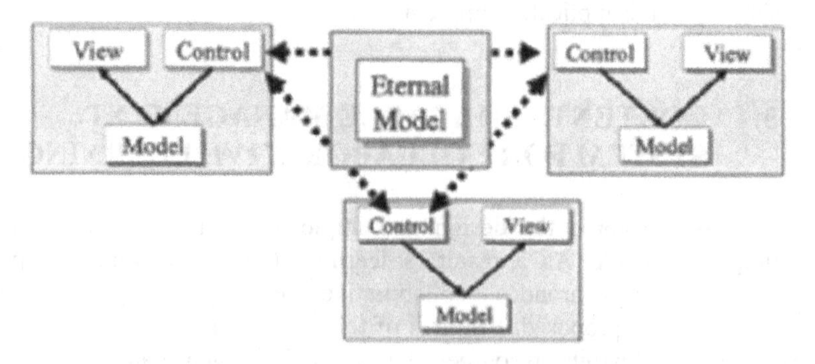

Figure 1. MVC·Architecture of the Proposed System

download of the shared contents from the server are done through Horb or HTTP. The user accounts are managed in the server. The user belongs to a group. The learning contents are shared between group members. The user freely accesses the shared contents of the group to which he belongs. The user makes his learning contents in his private workplace. This user is able to upload his learning contents to the collaborative workplace. In addition, the shared contents of group A are loaded in a learner's private workplace, and then this content is able to be loaded in the workplace of group B by that leaner. The Share object manages the shared contents. The group has multiple Share objects. The user who belongs to the group creates a Share object. Contents controlled in a Share object are stored in the Repository object. The Repository object controls not only the shared contents but also the relation information such as their attributes and the access permissions to them.

3.2 Design Concept

The Lapchat is a leaning tool to support preparation (information collection/arrangement) /open / shared / exchange of an electronic notebook in the ICT environment. This tool does not disturb the ICT environment which a learner daily utilizes for learning. Simultaneously, this tool offers a sophisticated mechanism which realizes uniform information management. Features of this tool are shown in the following:

- OS independence in of the client environment.
- The independent of the application utilized by the client terminal.
- Independent in the type of contents of a management object.

In the Lapchat, the method and process of information generation / discovery, and operation of the application which an individual learner utilizes are not shared. Especially, the following are important:

- Security/privacy by the learner terminal,
- Guarantee of the ownership for each learning contents.

Sharing of application operations at the Meta level has been realized. The individual who runs an application carries out the operation to that application. Then, these learners make open this result. In the Lapchat, the learner shares only the changed result.

3.3 Contents-Sharable Management

The Lapchat works in a client environment with a window interface (such as Windows, MacOS, Linux and so on). The image in the window that the learner designates is uniformly controlled as image information. In addition, postscript data (drawing and character, etc.) for the image information is controlled as the other image layer. By this, only the addition and deletion of the postscript data are enabled, in the condition that the image information was ensured in the lower layer. The management object by the Lapchat is shown in the following:

a) Contents image in a window interface on client desktop,
b) Image data in the client terminal,
c) A URL reachable image data,
d) Drawing and character string added to the data of a) ... c),
e) Postscript /annotation of a) … d). (generation day, creator, comments, title, etc. are contained) and
f) Discussion information for the a) … e).

This information management is done by a hierarchical log data management method (Kayama & Okamoto 2002) which was proposed as a function of a collaborative learning framework (Kayama & Okamoto 2004).

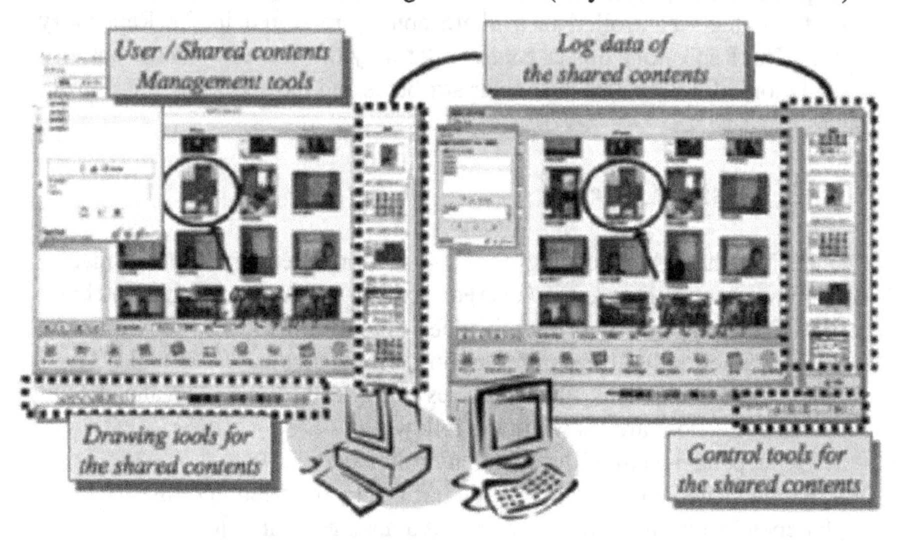

Figure 2: Interfaces of the Lapchat

3.4 Functions

The interface image of the proposed tool is shown in figure 2. For any type of learning (individual learning, collaborative learning or lecture) the following five functions are provided.

1. Persistence attribute of the collaborative workplace.
2. Linkage of the individual learning contents to the group learning contents.
3. Drawing for the shared contents,
4. Maintenance for the history log of the shared contents
5. Operation permission for the shared contents history.

3.4.1 Persistence attribute of the collaborative workplace

In this function, a user can set the externalization/ deletion permissions of shared contents which are formed in the collaborative workplace. These permissions are set by a dialog shown on the left top part of figure 2. The externalization of contents means leaving the learning result on the server after the learning activity is finished. By making contents into externalization, the reference and the additional change to the shared contents become possible, even if after the learner logouts from the collaborative workplace. When a user does not set the contents as externalized, the shared contents are automatically discarded at the end of the learning. The deletion permission means to decide the propriety of the shared contents deletion by the group member.

3.4.2 Linkage of individual contents to groups

The individual learning contents which are able to be uploaded to the collaborative workplace are as follows.

- Window image on the learner's desktop (includes the steal image for the dynamic application),
- URL reachable information,
- Image file in the learner's terminal,
- The previous shared contents.

To upload the individual contents to the collaborative workplace, the learner uses the right bottom part of the figure 2.

3.4.3 Drawing for shared contents

The drawing of the shared contents is carried out by the left bottom part of figure 2. In the Model, the shared contents and the drawing data for it are distinguished. Each of them separately managed. Therefore, it is possible to realize renewal only of the shared contents or renewal only of the drawing data.

3.4.4 Shared contents log management

The log data of the shared contents with their drawing data is retained in the Model. The history data is made to be the shared contents again. Then loading the current/past shared contents by individual learner to his private workplace is possible.

3.4.5 Operation permission for the shared contents history

Operation permission for the shared contents history is as follows:

- Addition of shared contents to the history list,
- Uploading of the other history data to the collaborative workplace,

- Modification of uploaded data,
- Deletion of the history data.

Group management is controlled by accessibility to the shared contents on the server.

3.5 LEARNING with the Lapchat

An outline of learning by Lapchat is shown in figure 3.

3.5.1 Individual Learning

The Lapchat client is started as a local connection (local://), and it is made to be the individual learning mode. In this mode, the 1st learner constructs his electronic notebook. Five kinds of information which are shown in (a), b), ... e) are controlled by repository for the individual (PR:Personal Repository). PR is constructed in the local directory (Removable Disk, Hard Disk, Floppy Disk, etc.), which the learner designated. PR is transportable. Then, with the Lapchat client, a learner can continue his learning without choosing the learning environment.

3.5.2 Collaborative Learning

It becomes the collaborative learning mode, when the Lapchat client is started in server connection (laph://). Synchronous collaborative learning is possible when other learners login to the Lapchat server at the same time. Equivalent permission to access the shared contents is given to all learners. Each operation in the identical timing has been realized. Asynchronous

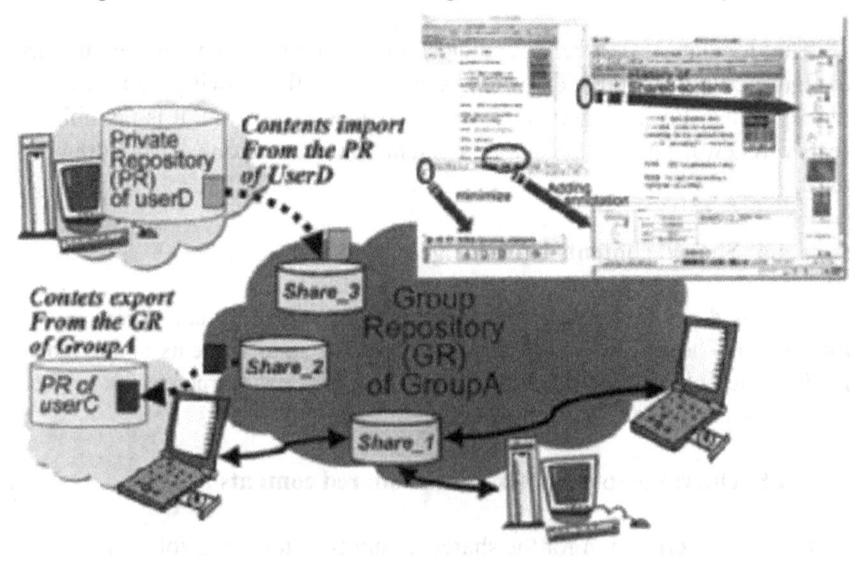

Figure 3 Outline of the learning with the Lapchat

learning will begin when another learner does not exist at the server in the same timing. In collaborative learning mode, six kinds of information which are shown in (a), b), ... f) are controlled by a repository on the Lapchat server for the group (GR:Group Repository). In this mode, the following actions become possible, whether synchronous/asynchronous learning type.

- Construction of the electronic notebook as a collaborative work in multiple learners (GR),
- Opening of the content of individual RP of the group member,
- Saving of information of GR to individual RP.

3.5.3 Lecture

A lecture is realized by controlling the shared contents in GR in the collaborative learning mode. The restriction for the shared contents means the following five kinds of setting.

1. Drawing permission to the base image of the lowerest layer,
2. Change permission of base image,
3. Addition permission of the history of the shared contents,
4. Deletion permission of the history of the shared contents,
5. Correction/change permission of the history data attributes.

The learner, who has a teacher role, makes the electronic notebook in his PR at the individual learning mode. Then, the learner opens to public his note from his PR to the GR. In the collaborative learning mode, the learner makes all shared restrictions OFF. When his group members participate in learning at the collaborative learning mode, the lecture is started. In the lecture situation it becomes possible that the learner preserves the shared contents, which the teacher presents in his PR. By writing annotations additionally to the preserved contents in his RP electronically note-taking is realized.

4. CONCLUSION

In this paper, the Lapchat, which supported learning in the ICT environment, was proposed. The Lapchat realizes the construction / open / shared of the electronic notebook. It is independent of OS and applications for learning in the client environment. Also, the concept of Contents-Sharable is proposed. A seamless linkage between contents in the private workplace and contents in the collaborative workplace is offered. Exploration of a group model based on the analysis of the postscript/ annotation information will be tried in future.

5. REFERENCE

O'Malley, C. (Ed.) (1994) Computer Supported Collaborative Learning, Springer-Verlag.

Kayama, M. & Okamoto, T. (2002) Collaborative learning in the internet learning space, International Journal of Industry and Higher Education, IEEE, August, pp.1-11.

Kayama, M. & Okamoto, T (2004). PLATFORM AND FUNCTIONAL MODEL FOR COLLABORA- TIVE LEARNING, Advanced Technology for Learning, Vol.1, No.3, pp.139-146.

Suthers, D. (2001). Architectures for computer supported collaborative learning., IEEE International Conference on Advanced Learning Technologies, pp.6-8.

Jackson, L. (1999). Concurrent Engineering in Construction -Challenges for the New Millennium-, Proc. of the 2nd Int. Conf. on Concurrent Engineering in Construction, pp 37-46.

Zhao, J. & Hoppe, H.U. (1994). Supporting flexible communication in heterogeneous multi-user environments, Proc. of the 14th IEEE Int. Conf. on Distributed Computing Systems, pp.422-229.

Goodman, B., Geier, M., Haverty, L., Linton, F. & McCready, R. (2001). A Framework for Asynchronous Collaborative Learning and Problem Solving, http://downloads .openchannelsoftware .org /SAILE/SAILE2001.pdf.

Plaisant, Rose, A., Rubloff, Salter & Shneiderman (1999). The Design of History Mechanisms and Their Use in Collaborative Educational Simulations, Proc. of the 3rd Int. Conf. on Computer Support for Collaborative Learning, pp.348-359.

Bereiter (1997). Situated cognition and how to overcome it, In D. Kirshner & J.A. Whitson (Eds.) Situated cognition: Social, semiotic, and psychological perspectives, pp.281-300.

M.A.Constantino-gonzalez, Suthers & Santos (2003). Coaching Web-based Collaborative Learning based on Problem Solution Differences and Participation, Int. J. of AIED, Vol.13, pp. 263 - 299.

SUPPORTING TEACHERS' PROFESSIONAL DEVELOPMENT THROUGH ICT

Reflections on two case studies

Leonard R Newton
School of Education, The University of Nottingham, UK

Abstract: This paper discusses the potential of two information and communications technology tools designed to support the management of school teachers' professional knowledge. Framed by drawing on experience of teacher education in England, the paper is structured around a consideration of what is meant by teachers' professional knowledge and issues in technology-mediated professional development. Two evaluative case studies of ICT-mediated knowledge management for teacher development purposes are presented. First, an evaluation of a curriculum development project in initial teacher education in several European countries; second, a project to develop more experienced teachers' skills in the teaching of aspects of enquiry-based science in UK secondary schools. Implications of innovative approaches to the management of teachers' professional knowledge using ICT are discussed.

Keywords: Knowledge management, knowledge for teaching, digital video.

1. INTRODUCTION AND CONTEXT

Recent rends in teacher education in England have focused on models that see teacher education as the development of a set of competences. Simultaneously, centralised control of teacher education curricula and accountability have enabled diversification of routes into teaching and a shift from the traditional role of universities and teacher education colleges, towards more school-based training models.

Arguably, the shift to a more standards-driven and school-based agenda for instruction of student teachers has coincided with a shift from the instructional process being an educative one towards a more training-orientated model. One, perhaps unintended, consequence of this trend has been that opportunities to develop understanding of the theoretical bases of teaching and learning have been eroded as notions of teaching as a technical

Please use the following format when citing this chapter:

Newton, L.R., 2007, in IFIP International Federation for Information Processing, Volume 230, Knowledge Management for Educational Innovation, eds. Tatnall, A., Okamoto, T., Visscher, A., (Boston: Springer), pp. 43–51.

process – a set of skills and behaviours to be modelled and developed in apprenticeship training - has emerged.

Emphasis on acquisition of skills at the expense of theory in teacher education may weaken teachers' knowledge of the theory underpinning practice. It may also diminish teachers' roles in reflecting on and researching their own practice. Such diminution threatens teacher autonomy and challenges teaching as a research activity. Significantly, it risks undermining the capacity of the profession to respond to change at a time when the pace and impact of innovations in education has probably never been greater.

Notions of the reflective practice (Schön 1987) have been at the core of ideas about the development of professionals across a range of settings for many years. Paradoxically, as the systems and structures of teacher education programmes have shifted to accommodate a greater element of school-based work, so the opportunities for beginning teachers to engage with a rich diversity of practice and to reflect together as novices and experts on theories of teaching have been squeezed. Thus teacher preparation models which involve singletons or small numbers of novices working in an equally small number of settings is likely to offer a less diverse and professionally rich basis for understanding linkages between theory and practice of teaching.

The following discussion considers two innovative tools that can support professional collaborative reflection through ICT mediated environments.

2. ORIENTATIONS IN TEACHERS' PROFESSIONAL KNOWLEDGE

It is helpful to consider teacher education as a process which seeks to build novices' professional knowledge for teaching. Therefore the subsequent discussion is informed by more explicit categories of knowledge for teaching proposed by Shulman (1987).

Table 1. Categories of knowledge for teaching (after Shulman 1987 p 8)

Knowledge of:	Amplification:
General pedagogy	General principles of classroom management and organisation
Pedagogical content	Understandings of subject teaching for learning

The two case studies discussed later in this paper draw largely on the knowledge areas of general pedagogy (Case 1) and Pedagogical Content Knowledge (Case 2) as detailed in Table 1. Although there is some overlap in each case between types of knowledge represented in Shulman's classification, the categories serve to disentangle the complexities of professional knowledge for the present discussion.

2.1 Orientations on knowledge in action

Schön's (1983) conceptions of reflective practice embrace the notions of reflection-*in*-action and reflection-*on*-action. The former can be described as 'thinking on you feet' (Infed, 2001) and the latter as thinking after the event. These notions remain pertinent to many aspects of contemporary teacher education, not least because the concept of reflection in action acknowledges that much of professional knowledge is exercised or 'situated' in complex contexts where there is a high degree of ambiguity (Pakman, 2000). Beginning teachers in England spend many hours in teaching practice and so, in principle, there is ample opportunity for development of skills of reflection-in-action. Schön also noted that professionals develop their own theories of action which could be articulated and critiqued in collective discussion (Pakman, 2000). This has been described by Schön as "*Reflection-ON-reflection in action*" (Schön 1987: 4). The importance of the situated and collaborative nature of professional learning has been highlighted by Lave and Wenger (1991) and more recently the close linkages between practice and theory have been conceptualized as 'praxeology' (Roth and Tobin, 2004). However in the context of technology-mediated knowledge management, some theoretical challenges have been raised over the locus of practice and learning in virtual communities of practice (Lueg, 2000).

From the perspective of education management, it is noteworthy that there has been considerable focus on system level developments in education (Fullan, 2000) and the arguments have been made for fostering a focus on collective learning for educational improvement (MacGilchrist et al, 1997), especially in turbulent educational environments. However, it needs to be born in mind that notions of organisational learning, whether at the level of the whole organisation or departmental level, are predicated on the learning of *individuals*. Therefore teacher professional development which emphasises individual learning can be viewed as a concomitant of collective learning at the department and school level.

The foregoing arguments serve to highlight the close interconnections between management of profession knowledge for personal professional development, school improvement agendas and collaborative learning.

3. IMPLICATIONS FOR SOFTWARE EVALUATION

Opportunities for collaborative professional reflection have the potential to build teachers' collective professional knowledge. ICT tools designed to support teacher development might therefore be expected to be deployed in situations that have at their core the opportunity for collaborative reflection. This can present a challenge for designers of ICT-mediated teacher development materials since software is often developed outside its context of use or considered for use by individuals working alone. Evaluation of educational software has become more refined, moving from simple descriptions of intrinsic features and usability issues to embrace consideration of end-use contexts. So evaluation tools have been developed which adopt a more situated approach that seeks to take account of the mode and context of use of software (Squires & McDougall 1996; Squires and Preece 1996). This alignment is useful in considering the potential impact of software for knowledge management.

4. CASE STUDIES IN KNOWLEDGE MANAGEMENT FOR TEACHER EDUCATION

This section describes two examples of innovative approaches to development of teachers' professional knowledge. The first case study is a synopsis of an evaluation carried out by the author for a European Community funded project in teacher education. The second case study reports work in progress which is developing novel approaches to facilitating reflective discussions between teachers, on aspects of science education practice.

4.1 Case Study 1: The TICEC Project

This case study was an evaluation of a *Socrates Comenius 2.1* project: *Professional Training of High School Teachers*. The project led to the development materials for use in teacher education that focused on the use of case studies as the pedagogic tool, presented through the medium of ICT. The broad aim of the project was to '*develop a methodology which uses case studies along with IT to enhance initial teacher training.*' (http://ticec.cliro.unibo.it/). The outcome of the project was a suite of CD-ROM materials and a training conference.

The titles of the CD-ROM materials covered a range of curriculum contexts and classroom practice. Some topics were generic in nature and so had potential application across a range of specific subject disciplines; for example, 'beginning and ending a lesson' and 'teachers' questions'. Others, such as 'drama' had a stronger subject focus and were therefore likely to be of less value outside their specific curriculum areas.

As argued above, best practice in teacher education involves novice teachers in a carefully constructed mix of observation of exemplary practice, model teaching, and practical classroom experience; ideally these are theoretically framed and critically evaluated. As such these experiences serve to support novice's critical reflection on their own and others' professional practice. The richer the novice teachers' experience and exposure to these learning opportunities, the more successfully embedded are their professional knowledge, skills and understanding of teaching: this accords with Schön's views of professional knowledge set out earlier. In the TICEC project, the team produced sequences of digital video of teaching episodes and presented these integrated with commentary in the CD-ROM format. The design of the materials within each CD-ROM reflected the underlying pedagogy developed by the project team. Each CD-ROM typically comprised of the video sequences, supporting documents (for example lesson plans), discussion prompts, a glossary of terms and a bibliography. In addition there were reflection points for users' consideration that served to summarise issues raised in the materials and set up points for future consideration. The video episodes of teaching and classrooms provided a useful complimentary source of materials for novice teachers to reflect upon and to discuss with their peers and more experienced professionals.

Digital video sequences were contextualised through introductory video sequences or text-based commentary. These resources sat independently in the material but they were also linked to text-based commentary which served several purposes. For example: an English transcript of the lesson; prompts and questions to focus attention on 'critical incidents' in the video and to promote discussion and reflection. CD-ROM material included commentary on salient professional issues which users could access after having considered the issues themselves.

The material tended to focus on the most practical aspects of teaching which was broadly appropriate as student teachers typically need to be able to quickly learn managerial routines and teacher behaviours that accelerate their transition into classroom practitioners. The CD-ROM products provided a wealth of exemplifications of teaching practice to provoke and encourage reflective discussion. Not all of the examples in the video sequences were necessarily of high quality practice, but that could serve as a useful aid to discussion with novice teachers, providing that it is carefully mediated by an experienced professional.

The technical-rational approach to development of professional knowledge may focus too heavily on acquisition of a narrow set of skills. The materials developed for the TICEC Project risked an over-emphasis on aspects of general pedagogical knowledge at the expense of other types of professional knowledge. This highlights the tension between the need to offer practical strategies to novice teachers whilst encouraging them to recognise that these strategies have underlying theoretical rationales. This risks losing an opportunity to develop teachers' critical thinking about efficacy and rationale or, in Schön's (1987) terms, reflections *on* action. It would be possible to use the TICEC materials in a very technically focused way, but one strength of the materials was that the theoretical basis of video extracts was acknowledged through the inclusion of bibliographies. Thus the potential was there for users to follow up areas of interest or need in the literature to develop a more theoretically grounded and evidenced-based approach.

4.2 Case Study 2: Developing Ideas and Evidence using 'ICE'

Here, a project to explore science teachers' use of digital video in professional development (PD) activities is described. Extensive use is made of video materials as a part of science teachers' professional development in England but there is a need for published research on its contribution and efficacy.

The project involved a team of science teachers and researchers in critically evaluating examples of digital video produced for PD purposes. The material used was selected from those produced for the IDEAS Project (2002); (Osborne, J., Erduran, S. and Simon, S. 2004) to support the use of argument in developing pupils' understanding of scientific ideas and evidence. The study made use of a novel software tool, the Interactive Classroom Explorer (ICE) to mediate PD activities and to facilitate the evaluative processes. The project set out to harness the professional knowledge of teachers involved in seeking to develop their practice in collaboration with others. Broadly the project goals were to identify the way in which teachers developed their thinking through use of PD videos, and to examine the role of ICE in promoting such developments to improve teaching.

4.2.1 The ICE Instrument

Interactive Classroom Explorer (ICE) aims to present digital video for exploration and discussion (face-to-face, online or offline) in a way that offers much more flexibility than is available with traditional video. ICE offers a much broader suite of tools than traditionally available, enabling the user to:

- play back a video while scrolling through a "timeline" containing a transcript, lesson plan and/or other annotations;
- create 'video quotations' (i.e. short video clips) that can be pasted into emails or a discussion area;
- display as 'pop-ups' resources associated with the video – e.g. still images of whiteboards, student work, copies of worksheets etc., selected either from a menu or from hyperlinks in the time-line;
- work with a collection of tasks, videos, timelines, texts and other resources that form a "module" containing everything needed for a particular professional development or learning activity.

4.2.2 Project Rationale

The project sought to investigate the efficacy of features of the ICE tool in supporting teachers' collective reflections-on-action in a key aspect of contemporary science education in England, namely *Ideas and Evidence.*

The development of this work has been informed by the experience of using ICE in other PD contexts (Harrison et al 2006). In the research discussed here, a series of short video sequences in which teachers and pupils are seen in classroom contexts working on developing skills of argumentation in science lessons was selected. This approach to presentation of argumentation used in the video had an extensive research basis (see for example, Erduran et al, 2004) and the project materials used in the present study was a development of that research project (IDEAS Project, 2002).

When working with ICE, teachers were presented with video and a simultaneous scrolling transcript and timeline. Other windows in the ICE tool present information to users, offer a discussion area into which contributions can be posted and an 'assignments' area which was not used in the present study.

A small group of experienced science teachers was recruited to work with the researchers. Following project briefing and training in using ICE, the volunteer teachers were asked to view IDEAS video sequences and identify professionally interesting aspects of practice exemplified in videos. The teachers were free to comment as they wished on the material. They were asked to post contributions to the on-line discussion in response to the video and to other postings. They were asked to use tools within ICE to identify video clips (known as 'video quotations') which illustrated their postings.

The teachers' postings were subjected to a content analysis and follow up interviews were conducted. Postings were typically data-rich and rooted in personal professional practice. Many of the areas focused on for discussion related to issues of classroom management and organisation prompted by the video. Interestingly, there was an emphasis in the early stages on the hindrances to engagement with the material. Some of these appeared to be linked to the 'overheads' in acquiring proficiency in using the ICE interface but others reflected issues of participation in online communities of practice.

5. IMPLICATIONS FOR IMPLEMENTATION

The case studies outlined above offer novel approaches to managing aspects of teachers' professional knowledge for development purposes. The TICEC project focused primarily on novice teachers in training and the ICE/IDEAS project was directed at more experienced teachers. From this experience a number of inferences about the design approaches of the two projects can be drawn, which have implications for knowledge management using ICT in teacher development contexts.

First, both projects make use of ICT as a tool for disseminating professional development opportunities. In the case of the TICEC project these opportunities were mediated through CD-ROM material which is easily distributed and can be accessed by individuals or small groups in professional development settings. By contrast users interacted directly with the IDEAS material in ICE either offline or online modes. These features facilitate collaboration between users over distributed or local networks. One key difference between the two approaches was the nature of the community of practice that could be established around the tools. In the case of the TICEC project the design enabled the materials to be used in extant groups. By contrast, ICE was both the means of establishing and of mediating the distributed group. Use of distributed groups has the potential to support PD rolled out on a wide scale, which may support national initiatives' more flexibly. But this is not unproblematic (Lueg 2000). However, the potential advantages for PD of a distributed group operating within ICE have been shown likely to be contingent on developing protocols for establishing a functional virtual group.

Both case studies made use of a blended approach to presentation of material. They shared the inclusion of video sequences supported by other materials. In the case of the TICEC materials, any reflective discussion on the content is contingent on the relationship to other artefacts in the PD 'system' (e.g. a tutor). In the case of the ICE tool, discussions were an intrinsic feature of the software design. The discussion content itself becomes a part of the shared artefacts of the PD episode within ICE. Users have control over the extent and nature of their postings to the shared discussion but it is the software that acts as the vehicle for discussion and the users are in control of the discussion topics independently of a tutor. To that extent, the ICE tool offers the opportunity for users to identify and develop discussion on issues that are of immediate and situated professional relevance, rather than imposed from outside the immediate context of use. However the absence of a mediating tutor may constrain the extent of progression of the discussion.

Despite the design advantages of ICE over more self-contained PD resources, it is clear that the efficacy of the tool is very much dependent upon teachers' investing time and developing their ownership of using the tool. Such investment is not automatic even for motivated ICT users. Much appears to depend on the relationship between the design purposes of the knowledge management tool and their relationships to the broader contexts of use for PD purposes. Thus, although there are exciting future possibilities for ICE, there is scope for further research in the design and practice of such tools for the management and development of professional knowledge.

6. ACKNOWLEDGEMENTS

I wish to acknowledge the work of the TITEC Project team in making their CD-ROM products available to me for evaluation purposes. ICE was developed by Colin Harrison and Daniel Pead of the University of Nottingham School of Education, funded by the TEEP and SEP Gatsby Technical Education Projects, and programmed by Thanassis Tsintsifas and

Pavlos Symeonidis of Labyrinth IT Limited, project leader Colin Higgins. Acknowledgment is made for making the ICE tool available to the author.

7. REFERENCES

Erduran, S., Simon, S., & Osborne, J. (2004). TAPping into argumentation: Developments in the use of Toulmin's Argument Pattern in studying science discourse. *Science Education*, 88(6), pp.915-933.

Fullan, M. (2000) The return of large scale educational reform. *Journal of Educational Change 1*, 15 – 27.

Harrison, C., Pead, D. and Sheard, M. (2006) "P, not-P and possibly Q": Literacy Teachers Learning from Digital Representations of the Classroom. In Michael C. McKenna, Linda D. Labbo, Ronald D. Kieffer and David Reinking (eds.), *International Handbook Of Literacy And Technology*. Mahwah, NJ: Lawrence Erlbaum Associates. Pp. 257-272.

IDEAS Project 2002 *Ideas, Evidence and Argument in Science Education* (IDEAS) Project details available at: http://www.kcl.ac.uk/depsta/education/ideas.html [accessed 11/10/05]

Infed (2001) *Donald Schon (Schön): learning, reflection and change.* Available at http://www.infed.org/thinkers/et-schon.htm [accessed 11 October 2005]

Lave, J. and Wenger, E. (1991): *Situated Learning: Legitimate Peripheral Participation.* Cambridge, UK: Cambridge University Press.

Lueg, C. (2000) *Where is the Action in Virtual Communities of Practice?* Proceedings of the Workshop Communication and Cooperation in Knowledge Communities at the D-CSCW 2000 German Computer-Supported Cooperative Work Conference "Verteiltes Arbeiten - Arbeit der Zukunft", September 12, 2000, Munich, Germany. http://www-staff.it.uts.edu.au/~lueg/papers/commdcscw00.pdf [accessed 12 March 2006]

MacGilchrist, B., Myers, K., and Reed, J. (1997) *The Intelligent School*, London: Paul Chapman.

Osborne, J., Erduran, S. and Simon, S. (2004) Ideas, Evidence and Argument in Science (IDEAS) Project, London: King's College

Packman, M. (2000) 'Thematic Foreword: Reflective Practices: The Legacy of Donald Schön', *Cybernetics & Human Knowing,* Vol.7, no.2-3, 2000, pp. 5–8. http://www.imprint.co.uk/C&HK/vol7/Pakman_foreword.PDF [accessed 6 October 2005]

Schön, D. *Educating the Reflective Practitioner* Presentation to the 1987 meeting of the American Educational Research Association Washington, DC. http://educ.queensu.ca/~russellt/howteach/schon87.htm [accessed 6 October 2005]

Shulman, L. S. (1987) Knowledge and Teaching: Foundations of the New Reform *Harvard Educational Review* 57, 1 pp1-22

Squires, D. & McDougall, A. (1996) Software evaluation: a situated approach. *Journal of Computer Assisted Learning.* 12, 146-161.

Squires, D. & Preece, J. (1996) Usability and Learning: evaluating the potential of educational software. *Computers Education.* 27, 1, 15-22.

Roth, W.-M., & Tobin, K. (2004). Coteaching: From praxis to theory. *Teachers and Teaching: Theory and Practice*, 10, 161-179.

Use of ICT by Primary Teachers

The Situation in Taiwan (The Republic of China)

Ian Selwood and Fang-Kai Tang

The University of Birmingham, School of Education, Edgbaston, Birmingham, UK.

Abstract: The empirical research in this paper examines teachers' attitudes to, and uses of ICT for administration and management. Additionally teachers' views were sought on the factors that would influence their use of ICT, and the effects of their using ICT for administration and management. The primary teachers in the sample reported generally positive attitudes to all aspects of ICT use for administration and management and a wide range of uses, but managerial use was somewhat limited. Some concerns were expressed relating to the age of hardware and the level of technical support available, and although they were generally happy with the quality of the training they had received, a very large majority felt they needed more training to improve further their skills and use of ICT for administration and management. Teachers noted improvements in communications and access to a variety of data since ICT was introduced for administration and management in their schools. Ongoing investment is needed if the situation is to be sustained and improved.

Key words: ICT, Primary Teachers, Administration, Management.

1. BACKGROUND

Taiwan, or The Republic of China, lies off the southeast coast of The Peoples Republic of China (mainland China). With a total land area of 36,179 square kilometres, and a population of 22.6 million in 2003, Taiwan is a densely populated country. Nearly 70% of the population live in cities where the density of population in the most crowded city is 9,826 people per square kilometre. Mandarin Chinese is the official language and is used by most people in their daily lives. Taiwan is a wealthy country with a substantial trade surplus, and foreign reserves are the worlds' third largest. The GDP in 2003 was 13,167 US dollars (Government Information Office, 2005).

In Taiwan, the Ministry of Education (MoE) has overall responsibility for the education system and Local Educational Authorities (LEAs) in the different regions administer regional education matters (Ministry of

Please use the following format when citing this chapter:

Selwood, I. and Tang, F.-K., 2007, in IFIP International Federation for Information Processing, Volume 230, Knowledge Management for Educational Innovation, eds. Tatnall, A., Okamoto, T., Visscher, A., (Boston: Springer), pp. 53–60.

Education 2005). Compulsory education in Taiwan currently lasts for 9 years. Six years in primary school (age 7-12) and 3 years in junior high school (age 13-15). According to the MoE, there were 241,118 pupils in 2,627 primary schools in Taiwan in the 2002-2003 academic year.

The Taiwanese government has implemented ICT-related policies in education since the late 1980s and this has included the use of Information Technology in Educational Management (ITEM). In 1998, the Ministry of Education budgeted 6.47 billion NTD (about 196 million USD) to carry out "The Plan of Information Technology Infrastructure in Education". (Computer Centre of Ministry of Education 2001). Under this plan, computer rooms were set up in all primary and junior high schools; training courses to improve teachers' ICT capability were provided; and educational ICT Recourses Centres were set up in every region in Taiwan to support and promote ICT in education.

In 2000 the MoE asked an IT company to design a system, the FES system, for governmental document exchange and this went live in 2001. By May of 2002, 51.2% of governmental documents were issued and exchanged by the system. Also in 2001, the MoE announced the "Master Plan of Information Technology in Primary and Junior High Schools" (Computer Centre of Ministry of Education 2001) to promote further ICT development in primary and junior high schools in Taiwan. In the section on educational administration and management in this plan, it stated that the use of ICT would simplify the process of educational administration and management and contribute to the connection between administration and management, and teaching and learning. Thus, the efficiency of administration and management could be improved, and teachers could also integrate ICT into their work and enrich the quality of teaching and learning. It should be noted here, that in schools, administration and management are not only the province of head teachers or senior managers, but also classroom teachers. (Selwood, Smith and Wisehart 2001). Teachers administer and manage in the classroom, in monitoring attendance, marking and assessing and monitoring pupils' progress (TTA, 1998).

To encourage and promote the use of ICT in educational administration and management in all regions, in 2002 the MoE sent out "The data exchange standard of student learning assessment and academic background database, version 2.0" to all primary and junior high schools to standardise the exchange of data (Computer Centre of Ministry of Education 2005).

Although the Taiwanese government has implemented a number of ICT policies and plans since the 1980s, there has been no published research focusing on the area of ITEM in Taiwan.

2. RESEARCH QUESTIONS

The aim of this study was to investigate teachers' use of ICT for administration and management in Taiwanese primary schools and sought to identify the factors that may influence this use. The following research questions were formulated to guide the work:

1. How do teachers perceive their ability to use ICT for dealing with their administrative and managerial work?
2. How do teachers use ICT for dealing with their administrative and managerial work?
3. What are the factors that would influence teachers to use ICT for administration and management?
4. What changes have teachers noticed resulting from the use of ICT for administration and management in schools?

3. RESEARCH METHODS

Data collection was undertaken in four Taiwanese public primary schools. As location and size were considered factors that could influence the findings the schools were chosen to be representative of large and small schools, and schools in rural and urban areas. Questionnaires were developed based on the work of Thomas et. al. (2004) and Selwood (2005) and, included both open and closed questions. One hundred and sixty questionnaires were distributed, and 92 were completed and returned. A response rate of 57.5%. Of the respondents, 27.2% were male and 72.8% female. The ages of the respondents ranged from 24 to 52 years old, with an average of the age 37.4. The average teaching experience was 8.6 years, and 19.6% of the respondents had special responsibilities as well as being teachers.

4. DATA ANALYSIS AND RESULTS

For clarity, in this section, the data and results are presented in the sequence of research questions.

4.1 Teachers' perceptions of their ability to use ICT for administration and management

If ICT is to be used for administration and management, positive attitudes towards ICT are essential. Teachers' attitudes on using ICT for administration and management tasks were therefore investigated, and these were generally positive. Teachers felt using ICT was easy (77.1%), not boring (84.7%) and did not make them nervous (88%); and they were confident (83.7%) and happy (91.3%) to use it. They also felt that using ICT could enhance their work efficiency (90.2%), reduce their workload (79.3%) and support their administration and management (91.3%).

4.2 Teachers' use of ICT for administration and management

In the questionnaire, teachers were given a list of topics where ICT could be used, and asked to indicate which tasks were carried out using ICT. Surprisingly exam entries and results (94.6%) was used most widely. In addition, more than four fifths of the respondents used ICT for record keeping (89.1%), organizing resources (85.9%), clerical work (84.8%), material preparation (84.8%), and writing reports (82.6%). Strangely, monitoring student progress was quite low (21.7%), bearing in mind how high record keeping was ranked. Also low was using ICT for registration 16.3% but this was due to only one school using this application and this obviously influenced monitoring attendance (18.5%).

A factor that influences the use of ICT is the availability and quality of hardware. Respondents' views regarding the adequacy of hardware are discussed later in section 4.3.1. In this section, we present data concerning the hardware used by teachers in school and at home. Desktop PCs and printers were the most widely used in both locations, with desktops being used by 97.8% of respondents at school and 79.3% at home; similarly printers were used by 97.8% at school and 77.2% at home. Over half the respondents also reported using Internet equipment (71.7% and 57.6% respectively) and digital camera (80.4% and 63% respectively) at school and at home. This high use of the digital camera merits further investigation. On the low use end, few respondents used a PDA at school or at home, and unsurprisingly few used a 'digital projector' at home.

Whilst hardware is crucial to the use of ITEM, the type of software used probably gives us a better indication of how teachers use ITEM thus the respondents were asked what software they used for administration and management tasks at school and at home. Word processor, Internet explorer, and e-mail were unsurprisingly the most widely used packages both at school and at home all scoring 80% usage or above. Additionally, 67.4% respondents used the school information management system from school, but access from home was limited with only 23.9% of respondents reporting this. Use of Spreadsheets (80.4%), PowerPoint (69.6%) and FrontPage (62%) all scored highly for use in school.

In demonstrating teachers' use of ICT probably more important than either hardware or software is the frequency that they use ICT and the amount of time they spend using ICT. 57.6% of respondents reported daily use of ICT for administration and management in school and 42.4% respondents at home. When we further examined the data, it could be seen that 81.5% of teachers use ICT for administration and management at least one per week in school and 72.6% reporting this use of ICT at home (at least once per week).

The most common amount of time that the respondents spent using ICT for administration and management was between 2 and 3 hours (40.2%). The majority of respondents (60.9% in total) spent 3 hours or less each time they used ICT at school. At home, there were also similar percentage of respondents (60.8% in total) using the computer for the same amount of time.

4.3 Factors, which influence teachers to use ICT for administration and management

A large number of factors are liable to influence whether teachers are likely to use ICT for administration and management. Previous studies (Selwood 2005) have shown these include support, training, quality of hardware and software. In the open-ended questions respondents pointed out many factors that influenced their use of ICT for administration and including: teachers' needs for access to hardware, software, and immediate technical support; training; time; workload and financial support.

4.3.1 Teachers' needs for hardware, software, and technical support

The majority of respondents (57.6%) agreed that the hardware in the school met their needs, 69.6% that software met their needs and 71.8% that technical support met their needs. However, this left a substantial minority of respondents who felt that their needs were not being met (43% hardware, and approximately a third for software and technical support).

Open-ended questions revealed that respondents felt that the schools should increase the quantity of hardware, particularly printers and digital projectors because with limited hardware, it was not convenient for teachers to use ICT for their work. In addition, they felt that the quality of the hardware should also be improved because much of it was obsolete and unstable and this influenced teachers' use of ICT. There were also some complaints concerning the speed of the school network at busy times like the end of term when large numbers of marks had to be entered into the schools' information management system.

More than half of the respondents agreed that the software used in their schools was efficient (78.3%), stable (63.1%), and user friendly (75.0%). In addition, more than four fifths of the respondents agreed that it was easy to find required information, input, and correct information. Around 85% of respondents agreed that the information format matches their needs, and the information could be used for planning and decision-making.

4.3.2 Training

The respondents were generally very happy with training they had received; with 86.9% of respondents agreeing that the ICT training they had received was helpful, and 88% agreeing their ICT skills had improved after training. However, only 59.8% of respondents felt that they had been trained in all necessary ICT skills for their work, and 95.7% respondents in total agreed that they needed more training to improve their ICT skills further.

ICT training was seen as important to improve teachers' ICT capabilities, and increase their use of ICT for administration and management. In the open questions, 12 respondents pointed out that providing more useful/relevant training courses to teachers was important in developing teachers' use of ICT. It was also suggested that the training courses would not only improve the trainees' ICT abilities, but would allow the trainees to cascade the training to their colleagues in schools. However, teachers were concerned that they might need to spend their own time attending the training courses, and this would affect their willingness to attend the courses. Furthermore, there was some concern expressed that; courses did not always match the needs of participants.

4.3.3 Time, workload, and support

In the open-ended questions, time, workload, and staffing (for technical support) were also seen as important factors by teachers. Time and workload are obviously closely related. Seven respondents stated that they had to deal with lots of work with their classes, and therefore their time to use ICT for administration and management was limited. They felt they needed extra time to use ICT facilities and also faced potential ICT problems caused by obsolete and unstable systems. Two teachers also highlighted the problem that schools had limited staffing to maintain the ICT facilities and provide immediate support when teachers needed it. However, respondents noted that they could seek help from their colleagues but their colleagues might be too busy to help them.

4.3.4 Schools' budget (for ICT)

Some of the factors noted above do in fact come down to financial issues. In the open-ended questions, seven respondents mentioned that lack of financial support was an important issue for schools. Noting also that finances affected such issues as: upgrading hardware and software, providing immediate technical support, and increasing staffing to maintain the facilities. Schools' budgets are obviously limited, but it was noted that they needed to devote more money to ICT. It was recognised that it was difficult for the government to provide sufficient and regular financial support to every school for ICT, but that more could be done.

4.4 Changes noticed by teachers resulting from the use of ICT for administration and management

58 out of 92 respondents answered that they found some changes in the processes. The positive changes mentioned include:

- Data storage, correction, and transfer had become easier; in particular, it was now easier to deal with students' marks and this led to time saving.

- It was more convenient to access information or teaching materials via the Internet, this again saved time.
- Communication had become faster: the use of e-mail as a communication tool between teachers and parents or pupils.
- Less wastage of resources, particularly paper, because a lot of information could be published or distributed via the Internet.
- Managing and searching information had become easier.
- A lot of work was now computerised, and thus the quality of the output had improved and become easier to control.
- Teachers were inspired to use ICT and had become more active in developing their ICT skills.

Additionally, there were also negative effects noted, including:

- Time was seen as problematic, for example: teachers had to frequently update the information on their homepages; spend extra time developing ICT capabilities; deal with problems of ICT hardware and software.
- Security of data was seen as problematic.
- If teachers relied on ICT more and more and the ICT facility was out of order, what would happen?
- Health concerns regarding computer use including the possible bad effect on the eyes from prolonged computer use.

5. CONCLUSIONS

The primary teachers in the sample reported generally positive attitudes to ICT use for administration and management with over 75% positive responses to all the measures used for this. Large numbers of teachers reported using ICT for a wide range of administrative and managerial tasks Each of the following tasks achieved more than an 80% response rate – exam entries and results, record keeping, organising resources, clerical work material preparation and writing reports. Notably only just over 20% used ICT for the managerial tasks of monitoring student progress.

Hardware used was predominantly desktop PCs, printers and Internet equipment. With respect to what software teachers used, the Word processor, Internet explorer, and e-mail were unsurprisingly high (all scoring 80% usage or above). Additionally, use of the school information management system, Spreadsheets, PowerPoint and FrontPage all scored highly (greater than 60%). Patterns of usage revealed high usage at both school and home.

Teachers' attitudes to the quality and quantity of hardware, software and technical support were generally positive. Nevertheless, concerns were expressed relating to the age of hardware and the level of technical support available. With respect to training, respondents were again generally happy with quality of the training they had received but nearly 96% felt they needed more training to increase their skills.

With respect to changes, teachers had noticed that related to the use of ITEM, again teachers were generally positive noting improvements in ease of storage and access to data (pupil records), ease of access to teaching resources and faster and more effective communication.

To conclude, the use of ITEM by Taiwanese primary school teachers appears to be well established and well received. However, there were some concerns regarding the need for sustained investment in infrastructure and training if the use of ITEM is to grow and play its full part in school improvement.

6. REFERENCES

Computer Centre of Ministry of Education (2001) The Plan of Information Technology Infrastructure in Education. MOE, Republic of China. http://masterplan.educities.edu.tw/conference/total.shtml (Accessed 25th July 2005)

Computer Centre of Ministry of Education (2005) Educational Administration and Management Computerisation. MOE, Republic of China. http://www.edu.tw/EDU_WEB/EDU_MGT/MOECC/EDU3620001/rd/admcmp.htm?TYPE=1&UNITID=37&CATEGORYID=0&FILEID=32202 (Accessed 28th September 2005)

Teacher Training Agency. (1998). New Opportunities Fund. The use of ICT in subject teaching lottery-funded training. Expected outcomes. London: DfEE.

Selwood, I., Smith, D., Wisehart, J., (2001) Supporting UK teachers through the National Grid for Learning. In Nolan, P., Fung, A.C.W., & Brown, M.A. (Eds), Pathways to Institutional Improvement with Information Technology in Educational Management. p 159-171. Boston: Kluwer.

Selwood I.D. (2005). Primary School Teachers' use of ICT for Administration and Management in Tatnall, A., Visscher, A. and Osario, J. Information Technology and Educational Management in the Knowledge Society. New York, Springer. p.11-22 (ISBN 0-387-24044-6)

Thomas, H., Butt, G., Fielding, A., Foster, J., Gunter, H., Lance, A., Pilkington, R., Potts, L., Powers, S., Rayner, S., Rutherford, D., Selwood, I., and Szwed, C. (2004) The Evaluation of the Transforming the School Workforce Pathfinder Project. London: DfES, Research Report 541. www.dfespublications.gov.uk

Centrally and Wide-Area Integrated Management of School Administration/Academic Affairs

Using a web application browser version

Kayo Hirakawa
Apsis Corporation

Abstract: te@chernavi is an application for the secure management of personal information and students' records in school. It aims to support schools that are undergoing big changes brought by structural reforms of the Japanese society from the perspective of school administration, which is a key part of school management. It realizes also a secure system irrespective of platforms thanks to system development with open resources.

Keywords: te@chernavi, ASP, information security, students' personal records, open sources, JAVA.

1. INTRODUCTION

"te@chernavi" has been planned and developed by our company since the year 2000. We have developed a system of 100% native web application for the transaction of academic affairs that is a key operation for schools, and we are offering it as a solution to achieve a wide-area centrally integrated management.

This centrally integrated system has already been in operation for 8 high schools that are run by Fukuoka prefecture for 4 years which is the first case in the whole of Japan.

Now we are at the next stage and conducting a new experiment in Saga prefecture. As a joint project of industry-government-academia, transaction by active server pages (ASP) of school administration of prefectural high schools is under way. A server is installed at a private company, and input/output is done with Syn-client at school. Aims of this experiment are to make clear the effectiveness of the central control as to the "personal information security" at school and to realize the "students' personal records" by a uniform management. When its effectiveness becomes clear, the education board to take advantage of the school administration by ASP

Please use the following format when citing this chapter:

Hirakawa, K., 2007, in IFIP International Federation for Information Processing, Volume 230, Knowledge Management for Educational Innovation, eds. Tatnall, A., Okamoto, T., Visscher, A., (Boston: Springer), pp. 61–68.

at all public schools in Saga prefecture, and in future will most likely allocate a budget. At present, disintegrated control is done at most schools and the reality is far from the uniform management and the central control in their real meanings. We strongly hope that our vision will meet with approval and come into wide use.

2. MODEL PROPOSED BY "TE@CHERNAVI"

2.1 Materialization of the Wide-Area Integrated Management by Web

The important factors of the system are:

- Shared use of a Web server for the transaction of academic affairs by multiple schools (ASP Model)
- Use of open sources
- Integrated management of data and application by server
- The possibility of Internet connection where available
- Platform independent
- Licenses don't arise from clients' OS and tools, as the building is with open sources.
- The maintenance is done on the server by remote control.

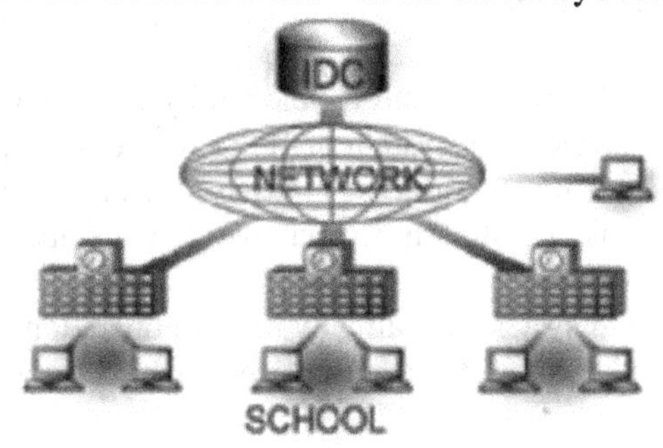

3. WIDE-AREA CENTRALIZED MANAGEMENT MODEL OF LOCAL GOVERNMENT

We have put into practice the first model of the centralized management in Fukuoka prefecture. Eight prefectural high schools of each different curriculum are managed with a server installed in a computer room at one of these high schools. Teachers at these schools start up a Web browser with a client on their individual desks to get access to the server directly and input scholastic performances, attendance, etc., or output ledger sheets.

The backbone of the network, "Fukuoka Gigabit Highway", is provided by Fukuoka Prefecture.

These autonomous networks use VPN (Virtual Private Network) for security.

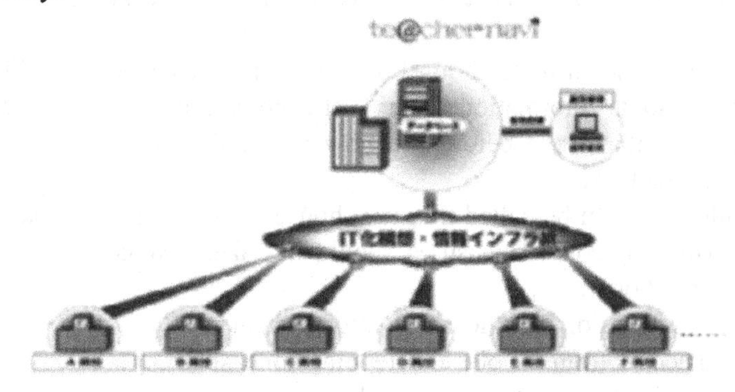

Features of this system are:

- There are no systems, no tools, no data at clients.
- Usable when the browser runs.
- Doesn't depend on the client's PC.

4. SYSTEM ARCHITECTURE

Features of the system architecture are:

- System construction of te@chernavi is done with open sources.
- OS is constructed with UNIX, Pure JAVA and it operates regardless of platform.
- Communication is done with the TCP/IP protocol.
- The environment includes any machine operating a Web browser.
- JAVA script is used for comfortable operation on the screen.
- Fine output of ledger sheets is through PDF.

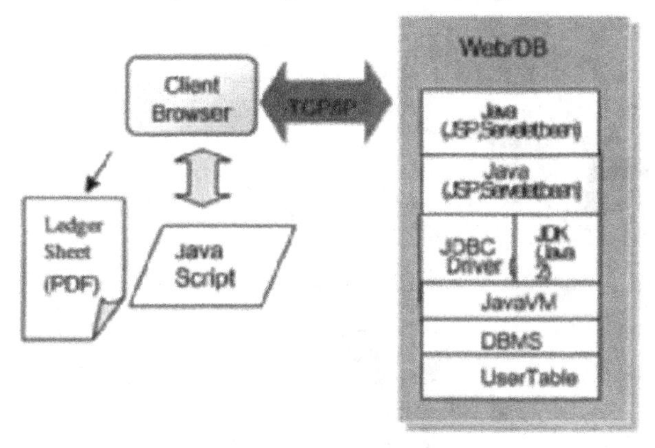

5. "ACT FOR PROTECTION OF COMPUTER PROCESSED PERSONAL DATA" AND PERSONAL DATA SECURITY AT SCHOOL

In Japan, the Personal Information Protection Law was enacted in 2005. Even for the personal information security at school, countermeasures that are much stricter than ever before are now required. Schools are under constant threat of attack.

Meanwhile, they can be called treasure- houses of personal information. Then, what kinds of countermeasures should be taken ? What are posing major threats to schools?

- Physical intrusion into the information system
- Illegal access from the outside(steppingstone, spoofing)
- Unauthorized access from the inside
- Computer virus infection
- Convulsion of nature(fire, damage by wind and flood, earthquake, thunderbolt)

The school is a treasure house of private information. Information on individual's privacy includes fields such as:

Address, Name, Telephone No., Parents' names, ... Academic record, Record of attendance, Special activities, Guidance record, ... Info on wishes as to the future course, Info on the future course ..... Address, Name, Age, Family makeup ..... Career history, Health check-up ..

Distributed Management at School Site involves:

- Final output/storing is in the form of "paper."
- Preparation process is with the "digital data."

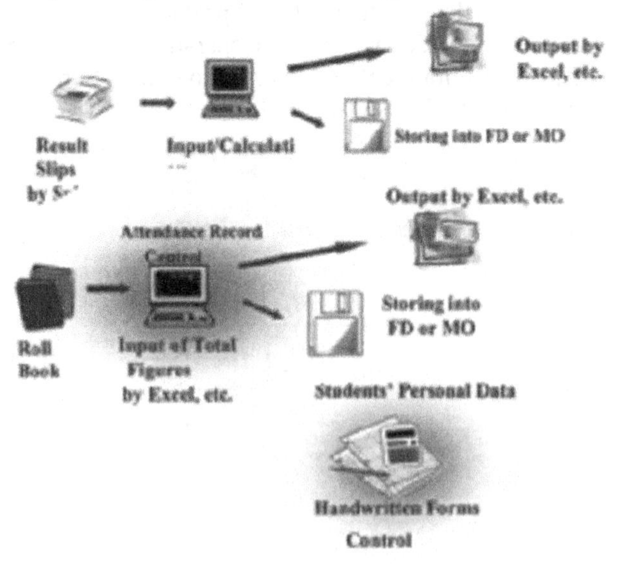

6. CONVENTIONAL ISSUES OF SYSTEM OPERATION AT SCHOOL

The situation in schools is characterised by:

- There are many handmade programs (Excel, Access, etc.)
- Each school has established its own system.
- There are transfers of teachers who are involved in the system operation and maintenance.
- A system and network controller is needed for each school.
- The required skill level is going up for the system controller.
- The backup work is complicated.
- The management/operation cost continues to increase.
- There is a risk of information leakage.
- The data dispersion makes it difficult to get the up-to-the-minute information.
- There is concern over possible inconsistency of data.
- Information is shared within narrow limits.
  Available machines are limited.

7. "TE@CHERNAVI" SOLUTIONS TO ISSUES

At School, materialization is by a centrally integrated management on-campus LAN server. At Education Board, materialization is by a wide-area centrally integrated management info server. This creates a system where:

- Real-time alterations/changes have become possible thanks to the remote maintenance.
- Trouble shooting has been speeded up thanks to remote maintenance.
- Security measures have been centralized and fortified.
- Dispersed management at the side of the client has been eliminated.
- Licenses don't arise from OS and tools.
- It doesn't matter what terminal machines and OS clients have.
- It doesn't matter what printing machines they have.
- The operation screen display has been made common.
- The cost of management and operation has been lowered.
- Consistency of data is guaranteed.

8. DEVELOPMENT OF SYSTEMS FOR SCHOOLS

In Japan, the government implemented e-Japan Strategy with the aim of becoming an IT nation from 2000 through 2005. In the field of education, improvement of infrastructure at public high schools was promoted nationwide. From 2006 onward, the government has launched u-Japan Strategy with the aim of becoming a ubiquitous society in which connection to the internet and its utilization will become possible “whenever, wherever, whoever, and whatever.”

Regarding the personal information security at school, there had been problems such as taking-out of information from office, loss, theft, and so forth in the era of stand-alone PC. The situation has become more serious these days because all schools have been connected to the Internet.

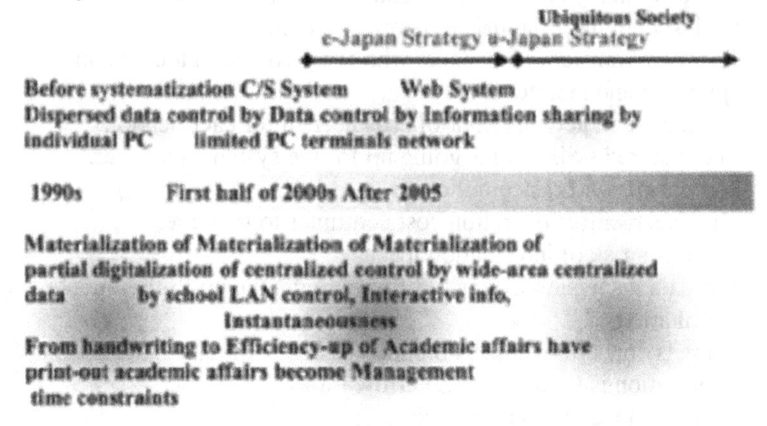

9. MATERIALIZATION OF “INDIVIDUAL CARTE”

“te@chernavi” has materialized the individual carte by means of a unified control of all personal information starting from entrance to school until graduation. The integrated and unified management has made it possible to look through information on each individual student, supporting teachers for meticulous guidance to their students. This sort of information can be integrated and outputted.

9.1 An Example

- Personal info (name, facial portrait, address, alma mater, etc)
- HR history while in school (academic year, class, ID no., home room teacher, etc)
- Attendance record while in school (absence, tardiness, leaving school early, etc.)
- School register changes (moving in, moving out, leave of absence, suspension, overseas education, etc.)
- Extracurricular activities, etc. (signing-up date, quitting date, etc.)
- Academic record history
- Health record (health data, utilization of the school nurse’s office, etc.)
- Guidance history
- Qualification obtained
- Course desired

10. SYSTEM CORRESPONDING TO JABEE

As one of its applications, we can refer to JABEE, that is, the Accreditation System for Engineering Education in Japan. JABEE demands standardization of criteria for the scholastic achievement assessment, credit earning and also disclosure of information. To this end, all elements such as attendance hours and academic results are to be translated into numerical points and its process needs to be verifiable rationally and objectively. te@chernavi is the system that meets these demands.

- Every-hour input of attendance (attendance/absence, results, tardy, reasons for early leaving, etc)
- Academic result data (raw scores, item scores, evaluation scores, ratings, etc)
- Grip on the process

JABEE demands preservation of the examination papers and various kinds of materials. "te@chernavi" can output all forms in PDF.

11. THE POSSIBILITY OF AN ACADEMIC AFFAIRS TRANSACTION SYSTEM WHICH ENABLES INTERACTIVITY AND INSTANTANEOUSNESS BY USING THE WEB

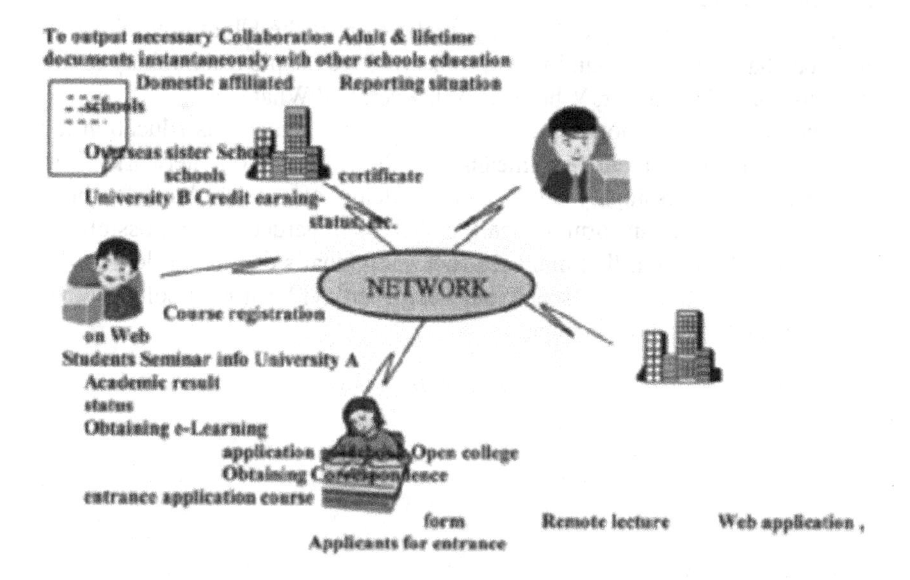

Finally, I would like to introduce a future image envisioned by te@chernavi. Thanks to the network that links school and students, a form of study irrespective of time and space will be realized. te@chernavi will provide a key administration processing service for the academic result control, accreditation, issuance of certificates, and so forth with its own school administration data-base. We believe that we should aim at realization of a meticulous education for each individual student by means of the information network. It is impossible for the conventional processing system of the academic affairs that was built in the age of uniform education to evolve into the system which manages each individual's data irrespective of time and space.

In this sense, our "te@chernavi" can be a solution and will contribute to the changes of schools. As a measure to counter violation of security such as hacking, setting up a transaction database is being considered separately from the main database.

12. CONCLUSION

Schools in Japan are currently at a big turning point. The dwindling number of children is forcing all schools to alter what a school should be. The centralized, uniformed and cram education method which has been consistently applied ever since the Meiji era after World War II is now under reassessment, and individualization and decentralization have become key words.

The advancement of computerization is expected to make a great contribution to this time for a change. Communication infrastructure at schools has been prepared under the government policy of the e-JAPAN Strategy (2000-2005.)

Since 2005, the u-Japan Strategy, realization of the Ubiquitous Society, has been hammered out, aiming at a society where people can hook up to the net and use it "Whenever, Wherever, Whoever, and Whatever."

I believe that we should aim at realization of a meticulous education for each individual student by means of the information network. The conventional processing system for the academic affairs that has been built in the age of the uniform education can be regarded as impossible to materialize the system that manages each individual's data regardless of the time and the place. In this sense, our "te@chernavi" can be a solution that will contribute to school change.

Communication Support Technologies for e-Learners

R. M. Bhatt
Garhwal University, Srinagar Uttranchal, India

Abstract: Advancements in telecommunication services and networking, have remarkably improved information communication services. The growing need of proper and secure communication is based on using appropriate communication technologies to work effectively and efficiently to meet the objectives of the information users. For some time this has been a great problem. For example, if one element e.g. a cable is supporting high bandwidth but a router limits it. In this paper supportive elements in information communication to enhance the communication performance are discussed. The technologies discussed here are: LAN/WAN, wireless, cabling Ethernet & ATM, iSCSI, IPv6 and outsourcing. These elements have been analysed, keeping in view how safely, efficiently, timely and exchangeable interfaces can be maintained for the sake of e-learners. These issues have been raised in the Indian context.

Keywords: LAN, WAN, iSCSI, wireless technology, Ethernet, ATM, outsourcing, IPv6.

1. THE CHALLENGE

Heavy investment is required to procure ICT items – this is the first and biggest challenge. Then the next challenge is to acquire the required items. My focus is to discuss this next critical challenge. Further, they should also ensure its secure storage and delivery. IT has provided great strength to any organization and innovation through IT has been earmarked for developmental work (Talero & Gandette 1995).

In answering this challenge, the emerging issues in support of information communication for integration of massive information across the different types of e-enabled users are discussed.

Please use the following format when citing this chapter:

Bhatt, R.M., 2007, in IFIP International Federation for Information Processing, Volume 230, Knowledge Management for Educational Innovation, eds. Tatnall, A., Okamoto, T., Visscher, A., (Boston: Springer), pp. 69–73.

2. THE INDIAN SCENARIO

The Indian population is 1.07 billion in 28 states and 7 UTs. The total literacy rate has grown to 65.04% (65.38% for males and 54.16% for women). The Indian elementary education system is the second largest systems in the world having an intake of 1.49 million children of age 6-14 and 0.12 million schools. For higher learning, there are about 320 state universities and over 15000 colleges. The total number of students in higher educational institutions was about 9228000 in 2005. Keeping this in view, in the fiscal year 2006-07 the education sector received 31.5% more funds than the last fiscal year (Education Times, Lucknow edition 2006).

More than 1000 pilot projects are being run by the government to spread IT among the masses. About $14.55 billion has been spent, with an estimated 40% success rate. Some, like Vidya Vahini, are providing connectivity to government senior secondary schools and Gyan Vahini is upgrading the IT infrastructure in higher learning institutions. The operation knowledge Government IT action plan (www.nic.in) envisages IT for all by 2008, computers & Internet access in all educational institution by 2003, and Government plans to create SMART schools and virtual institutions. Recently in a University communication the University Grants Commission (UGC) decided to establish "A UGC Network" named as UGCNET to provide a seamless, broadband, scalable nationwide inter-university link up to create virtual enhancement of the academic structure. Till now about 183 universities are covered under this net. In another attempt to strengthen the aim to disseminate education EDUSAT, an Indian educational satellite, was launched in 2004 to provide audio and video communication facilities. To increase the PC ratio the Government of India IT task force (Yojna 2000) has envisaged making a ratio of 1 PC per 50 people by the year 2008.

As far as the Internet economy is concerned, it is reported (Manjar, Rao and Ahmad 2001) that $22 billion is floating in India and 184 companies have invested in this economy. Further, 41% of companies have invested in the IT segment. These are some good indications.

3. LAN AND WAN

The basic elements for the formation of LANs were previously considered to consist of Network Interface Cards (NIC), hubs, switches and routers. Now integrated motherboards with NICs has reduced the number of element of the LAN. Further, due to a fall in prices and minimizing communication errors, switches are replacing hubs. Routers do not have much market and had a low growth (Pasha 2002) of 3.6% during the 2001-02. Now, the switching element has taken the lead and to a growth level of 28% during the 2001-02. IEEE has set the 802.1Q standard to enable VLAN information across a network and to enables timely delivery of highly delay-sensitive traffic, a signalling network priority on a per frame basis is required. For this purpose, a method has been provided under the IEEE 802.1P standard (Conver 1999).

4. WIRELESS TECHNOLOGY

A survey conducted in India by Express Computer (Wireless Survey 2002) found that a large portion (70%) of Indian companies have not deployed any wireless solutions but 10% are considering implementation. Interestingly, 47% implementation has been achieved for net access and communication purposes and for wireless implementation 48% of companies showed interest. Now WiMAX is emerging due to its superior range and bandwidth which can transfer around 70 Mbps over a distance of 30 miles to thousands of e-enablers from a single base station. For the new emerging ad hoc wireless networks, a routing protocol, termed Associativity Based Routing (ABR) has been augmented (Toh, Delwar and Allen 2002).

5. CABLING

Towards enhancement of the capacity of the cables, migration is on the eve of upgrade. For example, upgrading of Cat5 to Cat6 cable enhances capacity from 100 Mbps to 1 Gbps with a speed of 1000 MHz, though this is a costly affair. Fibre is too costly, so structured cabling is the best choice left among cabling solutions (Patra 2002). Cat7 is a recent improvement and is expected to capture 0.4% of the global market by 2006 (Naik 2005). Further, even fibre cabling is expensive but it also works for up to 70 km distance. The future will see a combination of Ethernet and Fibre technologies. DWDM (dense wavelength division multiplexing) is also being considered where a 1.6 terabit per second optical transmission has been noticed.

6. ETHERNET AND ATM

Originally providing shared bandwidth of 10 Mbps, Ethernet is now providing 100 Mbps dedicated connections through Fast Ethernet. To meet the 1 Gbps speeds of Ethernet switches, routers are not competent enough to match this speed. That is where ATM switches come into consideration for their deployment under the WANs. By making bursting connections these share resources stochastically under ATM and statistical gain has been achieved which enhances services. A burst-level priority scheme (Naik 2005) is allocated to burst on-the-fly according to their priorities which alleviates the under utilization of the resources.

7. ISCSI

To overcome the cost of fibre channel networks (Wireless Survey 2002), iSCSI can be implemented on a Gigabit infrastructure, but iSCSI does not yet meet standards and does not secure communication over a WAN. It can send the data over IP networks as well as over fibre channel over IP (FCIP), and can still run over Ethernet networks but FCIP run with the fibre channel technology (www.iSCSI.com). This can be viewed as an advantage for the future. As iSCSI is changing the networked storage landscape, its market is estimated to jump over $5 billion by 2007 from $216 million in 2003.

8. IPV6

IPv6 is a new generation Internet protocol designed by IETF, and now it will be possible to move from about 4.3 billion IPv4 addresses to over 3.4 trillion, trillion, trillion, trillion. Because of features providing an automatic routing and network reconfiguration, IPv6 has shown its advantages over today's IP (Wireless Survey 2002). Currently, its replacement cost is high (www.IP6v.com). This transition would cost around $1 billion per year.

9. OUTSOURCING

Outsourcing, another emerging phenomenon has resulted in significantly decreasing expenditures of enterprises and in turn, they gain high growth in marketing. Under this scenario advanced companies are outsourcing the software development jobs to different countries. Carmel (2002) recently pointed out that having human resources in abundance in India, to manage a global software project a virtual team could be developed. He categorized the software exporting nations into 4 tiers as Traditional (USA, EU, and Japan), 1st tier (Israel, India, Ireland), 2nd tier (Philippines, Russia, China), 3rd tier (Mexico, Romania, Pakistan, Costa Rica) and 4th tier (South Africa, Jordan, Bangladesh, Cuba). India has large firms and large projects which are the main reason for India's survival for a long period. It is known from discussions that in regard to export India is gaining $6 billion per year in software services. It was reported (Sanghi and Kirpalan 2002) that from 2001 to 2005 the American market would grow by 60%. In 2006 it is expected that $23.4 billion export revenue would grow by 27% annually. India has an opportunity of $100 billion to come (Carralho and Babu 2006).

10. DISCUSSION

According to NASSCOM (Express Computer 2003), a rise of 24% in IT spending (hardware, software & imports) in India is seen at about $15.5

billion by 2003. In the software and service sector, export was $22.5 during 2005 and is expected to reach $57 billion by 2009 as India controls 44% of the global offsource outsourcing market for software and back-office services. There are only about 7 million Internet users in India (expected to go up to 35 million), whereas the United States has one-third of the total Internet population. Population connected to the Net has shown a poor growth of 7% (www.bizasia.com). After a decade, the Indian IT services industry has shown consistent annual growth of 50-60% (Yojna 2006).

11. CONCLUSION

The upcoming scenario of ICT and other IT endeavours requires proper maintenance of IT infrastructure to meet the demand of providing a guaranteed service level. Therefore the discussed support communication elements should be properly deployed for e-enablers.

12. REFERENCES

Talero & Gandette (1995), Harnessing information for development: A proposal for World Bank Group Vision and Strategy, IT for development, 6:145-188

Education Times (Lucknow edition) (2006), The Times of India, 13th March, 2006, p-1.

Yojna (2000), a publication of planning commission, Govt. of India, New Delhi, No.10, Jan.2000, pp 4-10

Osama Manjar, Madanmohan Rao and Tufail Ahmad (2001). The internet economy of India, published by Inomy Media Pvt. Ltd., New Delhi,, 2001, pp. 10-16.

Akhtar Pasha (2002) *Rapid growth in LAN switching, NICs and hubs take a hit*, Express Computer, Vol.13, No.43, p. 11, Dec.30,2002, Mumbai

Conver, J. (1999) *Minding your virtual Ps & Qs*, Express Computer, July 26

Wireless Survey 2002, Express Computer, Vol.13, No.43, pp. 24,29, Dec.30,2002, Mumbai.

C.K.Toh, Minar Delwar and Donald Allen (2002) *Evaluating the Communication Performance of An Ad hoc Wireless Network,* IEEE Trans. on Wireless Commu.Vol.1,No.3, July,2002, p. 402-414

Gaurav Patra (2002), *Cabling: Looking a head*, Express Computer, Vol.13, No.43, p.34,Dec.30,2002, Mumbai.

Sushmita Naik (2005), *Cat7 : Waiting in the wings*, Express Computer, Vol.16, No. 21, p.19.

Fernadez, J. R., Mutka, M. W. (1999) *A burst-level priority scheme for bursty traffic in ATM networks*, Computer Networks, 31, 33-45

Erran Carmel (2002) *Global Software Outsourcing from Developing Countries*, Proceedings of the 7th international working conf. on Information & Communication Technologies and Development : New Opportunities, Perspective & Challenges, p.681, Ed. S.Krishna and Shirin Madon, IIM Bangalore May 29-31,2002, ISBN 1-901475-0-1

Sharad Sanghi and Karan Kirpalan (2002) *Why outsourcing makes sense,* Express Computer, Vol.13, No.43, p.30, Dec.30, Mumbai

Brian Carralho and Venkatesa Babu (2006) *Indian IT's $100 billion Opportunity*, Business Today, Vol.15, No.6, p.57.

Express Computer (2003) Vol.3, No.44, Jan 6.

Indo-US knowledge trade initiative (excerpts from FICCI Report on Knowledge Trade Initiative), Yojna, Feb.,2006,New Delhi, ISSN-0971-8400,P.15.

Using Educational Management Systems to Enhance Teaching and Learning in the Classroom

An investigative case study

Christopher Tatnall and Arthur Tatnall

[1] *Eltham North Primary School and Latrobe University,* [2] *Centre for International Corporate Governance Research, Victoria University, Australia*

Abstract: All schools store a considerable amount of administrative data on their students, and this is collected from a number of different sources. Depending on the requirements of their national or state education systems most schools store this data in a prescribed format so that it can be used to prepare reports for these education authorities and for local school use. The question we ask is whether any other classroom use can be made of all this data, and that is the question addressed in this paper. In the paper we examine a case study of a primary school in the state of Victoria, Australia and how it is attempting to come to grips with this question.

Keywords: Educational Management Systems, tracking and monitoring of student learning, enhancement of teaching and learning, integrated databases.

1. INTRODUCTION

Every school needs to store a large amount of administrative data relating to individual students and student cohorts. This data is collected from many formal and informal sources including: Student Enrolments, Early Years Interviews, Observational Surveys, Running Records, other Formal Testing and Anecdotal Notes. In Government schools in Victoria (Australia) the principal computer system used for this purpose is known as CASES-21, the use of which is mandated by the Department of Education for school administrative purposes. Many schools also keep other student data either on paper or in other unlinked systems. This research project was the result of a period of 'Teacher Professional Leave' (by Christopher Tatnall) and involved an investigation of whether:

1. Existing electronic data could also be used in some way to enhance teaching and learning in the primary school classroom.

Please use the following format when citing this chapter:

Tatnall, C. and Tatnall, A., 2007, in IFIP International Federation for Information Processing, Volume 230, Knowledge Management for Educational Innovation, eds. Tatnall, A., Okamoto, T., Visscher, A., (Boston: Springer), pp. 75–82.

2. Other student data stored by schools could be linked to the data already stored in electronic format in CASES-21 in order to facilitate things such as assessment and reporting. This type of data is primarily used for monitoring and accountability at the school level and could, potentially, be useful to classroom teachers to identify individual student needs and to improve student outcomes.

This paper reports on an investigative study of how several Victorian schools make use of their educational management systems. Data was collected from the following medium to large primary schools in the Northern and Eastern suburbs of Melbourne: Eltham North, Apollo Parkways, Glen Katherine, Milgate and Doncaster Gardens, as well as Scotch College, a large independent school in Hawthorn, and the South Eastern Region Computer Training Centre (SERCT) of the Victorian Department of Education. Another goal of the project was to develop an integrated database that could be used by classroom teachers to track and monitor student learning at any point in time during the children's attendance at Eltham North Primary School.

2. ADMINISTRATIVE SYSTEMS IN VICTORIAN SCHOOLS

CASES-21 (Computerised Administrative System Environment in Schools) is used in Victorian Government Schools to support their needs for administrative and financial management. An earlier version of CASES has been in use since the late 1980s (Tatnall 1995). It was developed as a tool for overall school administration and as a means of reporting back from schools to the Department of Education. No consideration was given to its use in school classrooms either to support teacher administrative functions or to enhance teaching and learning (Tatnall 1995; Tatnall and Davey 2001; Davey and Tatnall 2003).

CASES-21 aims to provide school administrative support staff with secure access to data entry and reporting modules that supports school administration and finance functions. The Department of Education claims that it has been designed to be modified to meet evolving school business needs. It currently has two main modules (Department of Education 2005):

- An Administration Module to provide student administration support, including the facility to manage: student and family data; student pastoral data; student medical information; student attendance; student achievement; student discipline/welfare; accident and incident data; activities (including student excursions); school management information; basic timetabling; daily organisation; and school associations (e.g. Parents Club and School Council). Almost all Government schools are now using this module.

- A Finance and Local Payroll module that aims to assist schools to: create and receipt family and student invoices; manage family debtors, sundry debtors and creditors; manage the school's asset register; process and manage the school's local payroll; manage school finances and budgets; generate appropriate financial reports. Currently only a small number of schools are running CASES-21 Finance whilst the remainder use the older version in CASES.

The Department of Education has eleven CASES Training Centres located throughout Victoria to provide professional development and training in the use of CASES and related administrative applications. Those attending this training are predominantly School Support Officers (administrative support staff) and School Principals. The School Support Officers attend to learn how to use the CASES system, while the Principals attend in order to gain an understanding of how CASES can be used to support the management of their school. Very few classroom teachers attend the training sessions as CASES typically has not been used by the classroom teacher. There is also a training course on 'Seemless Views' to show how CASES-21 data can be linked to Microsoft Access and FileMaker databases, as well as other applications such as Microsoft Word and Excel.

The prime purpose of CASES-21 is to enable reporting from schools back to the Department of Education. For security reasons each Government School in Victoria has two distinct (unconnected) computer networks: an administrative office network running CASES-21, and a curriculum network for use by classroom teachers. The curriculum network is wireless enabled but not the administrative network. As CASES-21 runs *only* on the administrative network a classroom teacher wanting to access CASES-21 data must use a computer in the school administrative office that is connected to this network.

Other school administrative systems provided by the Department of Education include building maintenance and teacher employment, but these are not relevant to this paper. A multitude of other administrative computer systems, both commercial and locally written, are also use in schools. These include:

- Systems used for reporting student information to parents
    - Commercial school reporting packages used in a large number of schools.
    - Some schools write up student reports as Microsoft Word documents without using any additional software.
    - A few schools use software that has been custom-built for them
    - Other schools still send handwritten reports home to parents.
- Systems for handling student achievement data
    - CASES-21 has facilities for handling data of this type, but as the purpose of this system is to transmit this data back to the Department of Education there are no additional facilities for reporting to parents.

  - All schools are required to enter data into CASES-21, but teachers do not have access to the program, and administrative office staff usually perform this operation. Data is typically kept in different ways by each teacher in each school.
  - Commercial packages are not used very much at present, but some of these that prepare reports to parents also have some basic ways of storing achievement data.
  - Use is made of database management packages such as Microsoft Access and FileMaker in some schools.
  - CASES-21 allows data to be typed in, but at the moment there seems to be no easy way of importing data from other databases or software applications.
- Systems for tracking various data related to students, such as:
  - Achievement.
  - Special programs.
  - Visits from visiting professionals such as speech pathologists, psychologists, eye specialists etc.
  - Additional homework etc for high / low achievers, individual learning plans.

These systems must work across year levels and keep track of each student's progress for the entire period he or she is at the school. They should also be accessible to all teachers (CASES-21 is not). One problem is that schools have different processes for storing and tracking this data. In some case this is handwritten, it may be in folders in filing cabinets, it may be in simple unconnected databases or spreadsheets, or it could reside in basic commercial packages.

CASES-21 allows the export of data to other systems, but it does not allow data to be imported from these other systems. This means that if the same data is required for use in several different applications, unless it is entered *first* into CASES-21 and then exported to the other system, it must be retyped for use in CASES-21.

3. ELTHAM NORTH PRIMARY SCHOOL: VISION, CORE VALUES AND PHILOSOPHY

"Eltham North Primary School is a community united by a commitment to learning. It was founded in 1924 and is intensely proud of its educational achievement in serving the community of Eltham. We are located one kilometre north of the township of Eltham Central, a residential suburb renowned for its peaceful bush surroundings. At the commencement of 2003, the school population of 425 students included 46% girls and 54% boys. Of the current student body, 98% are Australian born and 2% of our students do not use English language as their first language at home." (Eltham North Primary School 2003).

In the School's Charter, the vision statement speaks about empowering students to help them develop communication skills and relationships that recognise and explore the "increasing responsibility they have for their own learning". It states that the school will provide an innovative and progressive learning environment that "engages, challenges and extends the learning capacity of all students". The core values of the school's learning community articulated in the Vision Statement involve building up strong foundations upon which the students can achieve, and support themselves and the well-being of others. It points out that these values are embodied in the school motto "Growing, Discovering, Creating Together" (Eltham North Primary School 2003).

Eltham North Primary School's Charter goes on to indicate that the school's teaching philosophy is based on a commitment to prepare students with the attributes and competencies needed to turn them into "active and responsible citizens in an ever-changing global society". It indicates that the school aims to encourage and provide "a caring atmosphere where the children feel emotionally, socially and physically secure", and makes use of reflective learning and problem solving so that negotiation and skill development can be used to create a process for insightful curriculum development (Eltham North Primary School 2003).

3.1 Eltham North's Student Information System

Eltham North Primary School's current system for handling student data and keeping copies of reports to parents etc. involves storing this data in a filing cabinet. Each classroom teacher keeps summaries of their student data in a large A3 folder with a file in a filing cabinet for each student. At the end of each year all of this data is collected by support staff and collated, ready for redistribution to this grade's next year teacher. This is a lengthy operation, and as part of this project we developed a Student Information System, which collected and stored a lot of this data electronically.

The school recently committed significant resources to the development of this new system, which has functions such as:

- Giving classroom teachers access to the name, phone number and emergency contact details (downloaded from CASES-21) for their students, without the need for them to go to the school office system.
- Providing a history of each child's schooling: grades, years, teachers (no achievement data as yet – this is to be added in the future) and involvement in special programs.
- Track (keep a record of) each child's individual learning plans.
- Link Microsoft Word reports to a database entry for each child.

As the school makes use of Apple Macintosh computers, the database was written using FileMaker Pro 8, and is stored on a central server on the curriculum network. (The reader will remember that the school has two different unconnected networks and so data on one cannot be accessed from the other). Each teacher in the school has access to the database via their own laptop using the school's wireless network. The database is password protected with different teachers have different access levels.

Teachers at the school have readily embraced the new system which has cut down their administrative workload and made relevant student information readily available to them. Eltham North Primary School is committed to extending the use of the database by purchasing more FileMaker licences and providing time and resources to make further additions to the database system.

4. ADMINISTRATIVE SYSTEMS IN OTHER VICTORIAN SCHOOLS

The study also investigated administrative databases used in other schools. As there is no single program or package recommended by the Department of Education many schools have created their own databases or purchased commercial packages. Part of this research included involvement in a working party based at Glen Catherine Primary School. The working party included teachers and administrators from six different schools in the North Eastern Suburbs of Melbourne. One of the objectives of this group was to find a package to both assist with the writing of student reports and also to store data on students. A plethora of programs have been written and available for schools to purchase that make report writing easier for the classroom teacher, however we found that there were no suitable packages for tracking and storing data on individual students.

Several schools have created their own systems. We viewed a number of these but did not find one that would be suitable for Eltham North Primary School. Many schools have used FileMaker Pro to create very simple databases, but none of those seen were fully relational. One school had lots of different databases which were not linked, meaning that teachers had to write the same information into several different databases. They were also not linked to the school's administrative system (CASES-21). Several schools used Microsoft Access but this has limited application in primary schools as many use Apple Macintosh Computers. FileMaker Pro was seen as a good compromise as it can be used on both platforms.

The working party at Glen Catherine, after viewing many different school databases and commercial packages, decided to recommend a commercial package for their schools that is used primarily to assist classroom teachers to write student reports to send home to parents. They found that there were no commercial packages readily available to track student's performance and academic achievement over time.

We also visited the Austin Hospital School, a Government school that services this hospital and also has outreach programs in other Government Schools. This school had a database created for them in FileMaker Pro. As the school's student population is constantly changing, reports need to be written on individual children on a regular basis. The school also needed a good method of tracking children's involvement in special programs, and their database system does this for them. Although it is not currently linked to the CASES-21 network, in the future it could be.

The SERCT Centre visited offers a service to schools in the South Eastern Region assisting them develop their own databases. These are mainly designed to assist classroom teachers in writing parents' reports. None of the databases we viewed there, however, were used to track and monitor student academic progress or involvement in additional programs.

5. CONCLUSION

CASES-21 was designed to support *school office* administrative applications with reporting back to the Department of Education. It was not intended for use by *classroom* teachers. CASES-21 can be used relatively easily to export data to be used in other systems, however it cannot easily import data from other systems. This has led many schools to having two or more systems in operation, often containing similar data. The other problem with CASES-21 is that it is only available from the administrative network in schools (unconnected to the curriculum network accessible to classroom teachers), making it very difficult for teachers to access.

This study identified many features in CASES-21 that could benefit the classroom teacher but most are related to a teacher's administrative functions such as organising camps and excursions, sending letters to parents and so on. The problem of classroom teacher access to this system has meant that very few teachers can currently make use of this data. It is not that this is forbidden, just difficult. This study concentrated on how student data could be used to assist the classroom teacher to improve student learning, but did not identify any use of this data directly related to teaching. Most uses related instead to general teacher administrative tasks.

The investigation also found a number of other issues that would need to be considered before extensive use can be made of this data. Privacy was one important issue raised. CASES-21 and other systems in schools contain a lot of personal information on students and their families, and on teachers in schools. Most systems investigated in this study had high levels of security, but this is an issue that individual schools will clearly need to consider very carefully. Another issue was that of data storage. Developing systems to store and track student data over their time at school is a great idea, however after just a few years this will have built up to a very large amount of collected data. Some of this may be viewed regularly but other data not at all. Schools are required to keep all data on a student until this student has finished secondary school (Year 12). Data archiving is thus something schools will need to look into.

6. REFERENCES

Davey, B. and Tatnall, A. (2003). Involving the Academic: A Test for Effective University ITEM Systems. Management of Education in the Information Age: The Role of ICT. Selwood, I., Fung A. C. W. and O'Mahony C. D. Assinippi Park, Massachusetts, Kluwer Academic Publishers / IFIP: 83-92.

Department of Education (2005). Information and Communications Technology: CASES21. Web page, Date accessed: October 2005, www.sofweb.vic.edu.au/ict/cases21

Eltham North Primary School (2003). Curriculum and Charter. Web, Date accessed: October 2005, http://www.elthamnorthps.vic.edu.au/c6.1_charter.htm

Tatnall, A. (1995). Information Technology and the Management of Victorian Schools - Providing Flexibility or Enabling Better Central Control? Information Technology in Educational Management. Barta, B. Z., Telem M. and Gev Y. London, Chapman & Hall: 99-108.

Tatnall, A. and Davey, B. (2001). Open ITEM Systems are Good ITEM Systems. Institutional Improvement through Information Technology in Educational Management. Nolan, P. Dordrecht, The Netherlands, Kluwer Academic Publishers: 59-69.

WBT Content for Geography and Geology using VRML

Goro Akagi, Koichi Anada, Youzou Miyadera, Miyuki Shimizu, Kensei Tsuchida, Takeo Yaku, Maya Yasui
Nihon University, Waseda University Senior High School, Tokyo Gkugei University, Nihon University, Toyo University, Nihon University, Nihon University

Abstract: In this paper we report on WBT content for geography and geology using VRML. We also propose an idea of WBT content for local area study that has not yet been implemented and discuss its effect from the viewpoint of knowledge management.

Keywords: e-Learning, WBT, geography, geology, VRML.

1. INTRODUCTION

From the National Curriculum Standards for lower secondary schools (http://www.mext.go.jp/b_menu/shuppan/sonota/990301/03122602.htm), an important purpose of geographical education is to train students to discover characteristics of their local areas by relating spatial data to the environment of the area and the activities of people living there. To achieve this purpose, it is quite important to give experiences of finding out scenes of areas to learners by using maps and statistical data, so atlases and statistical sourcebooks are often used as teaching aids in geographical education at junior high schools.

We assert that 3D Computer Graphics (3DCG) can be applied to geographical education, because 3DCG can provide a more intuitive understanding of maps than paper-based educational material, and they improve learners' motivation to study geography and geology. From such a point of view, we experimentally developed Web Based Training (WBT) content using 3DCGs for geography and geology.

In this paper, we give an overview of our WBT content and report high school teachers' evaluation of it. Our WBT content includes 3D landform maps written in Virtual Reality Modelling Language (VRML) to explain the

Please use the following format when citing this chapter:

Akagi, G., Koichi, A., Miryadera, Y., Shimizu, M., Tsuchida, K., Yaku, T. and Yasui, M., 2007, in IFIP International Federation for Information Processing, Volume 230, Knowledge Management for Educational Innovation, eds. Tatnall, A., Okamoto, T., Visscher, A., (Boston: Springer), pp. 83–88.

characteristics of landforms and topics in geography and geology. Here, a 3D landform map means a 3DCG of a landform.

In the next section, a preliminary a discussion of 3D landform map is provided. In Section 3, we give an overview of the experimental WBT content we made, and in Section 4, we propose an idea of WBT content for local area study that has not yet been implemented and discuss the effect of the content from the viewpoint of knowledge management. Section 5 is devoted to a brief evaluation of our WBT content. In the final section, we discuss conclusions and further research.

2. PRELIMINARY: 3D LANDFORM MAP

The experimental WBT content that we made up includes 3D landform maps, which are 3DCGs of landform maps. These maps are written in VRML and were generated from numerical altitude data with different mesh sizes (i.e., 50m, 250m, and 1km) published by the Geographical Survey Institute (http://www.gsi.go.jp/ENGLISH/index.html). We employed several programs that automatically produce VRML files of 3D landform maps from numerical altitude data, and were developed by ourselves (see Anada, Kobayashi, Tsuchida, Miyadera, Motohashi and Yaku 2004).

Learners can browse 3D landform maps by using common Web browsers (e.g., Internet Explorer, Netscape and Firefox) with VRML plug-ins (e.g. Cortona (http://www.parallelgraphics.com/products/cortona/)) interactively changing viewpoints in 3D landform maps by moving a mouse (or a pointer device).

3. WBT CONTENT

In this section, we give an overview of the experimental WBT content we made. Our WBT content consists of the following: "Indexed glossary of geography and geology", "3D Inou-zu" (Inou-zu is the old Japanese map drawn by Tadataka Inou in the early 19th century) and "History and 3D landform map". The number of files used for each kind of content is shown in Table 1.

3.1 Features of the WBT Content

The most important features of WBT content are the following: each item is written in HTML or Javascript, VRML, or JPEG, so it can be exhibited on the Web and browsed by using normal Web browsers (see Fig. 1); learners can interactively move 3D objects and viewpoints in 3D landform maps, so they can view one object from various points of view; each item is completely independent, so learners can freely choose among courses. We provided the last feature, because atlases and statistical sourcebooks are used in a similar way in junior high school geography classes.

Table 1. The number of files in each content (2005.10.31)

	HTML file	VRML file	Other image file
Indexed glossary	192	122	25
3D Inou-zu	24	39	26
History and 3D landform map	8	40	3
Frame and others	28	0	24
Total	252	201	78

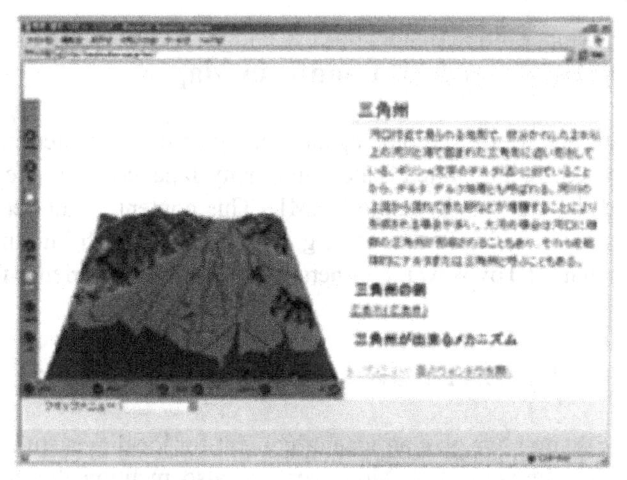

Figure 1. Screen shot of Internet Explorer displaying the WBT content (indexed glossary of geography)

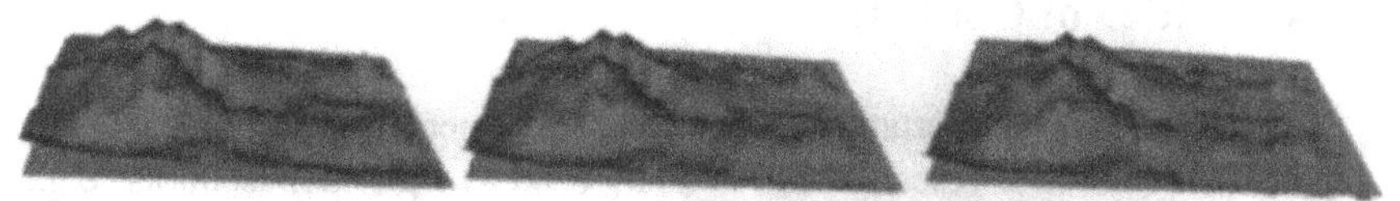

Figure 2. 3DCG explaining the mechanism of creating a ria coast. Learners can move the plane of the sea level by using a mouse.

The educational materials we made are regarded as WBT content because of the first feature. Furthermore, we also provided the second and third to achieve the purpose of geography education described in Section 1.

3.2 Indexed Glossary of Geography and Geology

Our indexed glossary of geography and geology included 81 words as of October 31, 2005. Each page includes one word and an explanation of it. We also made 3DCGs to help learners understand the explanations and attach them to the page. For example, a *ria coast* is a deeply embayed coast formed by a partial submergence of a landmass. To explain this, we made a simple 3D landform map of mountains and valleys with the plane of the sea level, and learners can interactively observe how the coast changes by moving the plane of the sea level (see Fig. 2).

3.3 3D Inou-Zu

We produced a 3D landform map and mounted textures of the Inou-zu, which is an old Japanese map drawn by Tadataka Inou in the early 19th century (Edo period), on it. This content helps learners to relate geography to history

3.4 History and 3D Landform Map

On Aug., 7, 1945, the old Japanese navy arsenals located in Toyokawa were bombed by the United States, and many Japanese were died there. We re-created the arsenals by using VRML. This content is not yet completed, and we plan to add information, e.g., the usage of each building, to a 3D landform map of Toyokawa. Learners will be able to experience the past.

4. CONTENT FOR LOCAL AREA STUDY

In this section, we give an idea of content for local area study, which has not yet been implemented. Moreover, we also mention the effects of the content from the viewpoint of knowledge management.

4.1 Idea of Content

Area studies are important in education for elementary school students in Japan. Students plan research and go out to areas near their schools, e.g., shopping arcade, farm, to collect information from people working there and view the information from various angles. A deeper understanding of their local areas results from such research experiences.To aid area studies, we propose 3D content, e.g., 3D virtual shopping arcades, 3D virtual high-rise buildings, and 3D virtual forests, that work as the interface of a 3D database. More precisely, learners can store information on their areas in the 3DCGs. We give a couple of example of such content in the following.

The usage of high-rise buildings: The usage of each floor in a high-rise building is correlated with the height of the floor, i.e., lower floors are used for shopping malls, banks, and so on, while, upper floors are used for offices and hotels, and the top floor is for restaurants and observatories. Learners find out about such correlation by arranging information on the usages of 3D virtual high-rise buildings.

Distribution of creatures in forests: Learners can find out the how creatures, i.e., plants, insects, and animals, are distributed in neighbouring forests by observing in forest areas and arranging data in a 3D virtual forest.

The content from learners can be shared over the Internet, so learners can understand the differences and similarities among areas (see Fig. 3).

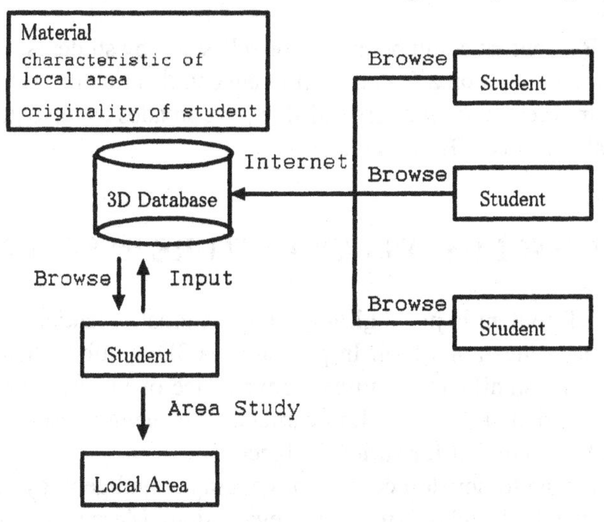

Figure 3 3D database and applications to area study

4.2 Effect of Content from the Viewpoint of KM

Knowledge Management (KM for short) has recently become important in education. According to Ogushi (2003), the previous educational method in Japan makes much of the stages of Combination and the Internalization, but not of Socialization and the Externalization of Nonaka's SECI model (Ogushi 2003); however, the stages of Socialization and Externalization play an important role in KM for educational innovations. We propose an application of our content described above to help local area studies at these two stages. In our content above, students have occasion to extract and show the characteristics of local areas, and such activities bring the Socialization of tacit knowledge and the Externalization of such tacit knowledge to become explicit. Furthermore, students can combine the explicit knowledge by comparing their works each other, and deepen understanding of the characteristics of local areas. This implies the Internalization of explicit knowledge. Therefore, the content for local area study provides cycles that appeared in the SECI model.

5. EVALUATION

We asked a couple of teachers at Waseda University Senior High School to evaluate our WBT content. Some of their comments are as follows:

a) Usually, pictures of landforms are used in geography class. 3D landform maps provide views from more sides and provide more intuitive understanding of topographies to students than pictures can.

b) 3D landform maps could also be applicable to studies on the urbanization of rural areas.
c) The WBT content could be useful for self-study by students.
d) The measurement on a 3D landform map could enrich the WBT content, and metro areas and underground shopping arcades are also good materials for this WBT content.

6. CONCLUSIONS AND FUTURE RESEARCH

Our WBT content is just a glimpse into the great potential of multimedia content as an educational tool. In particular, a 3D landform map written in VRML has the significant features: convenience of viewing, operation and creation, and rich interaction for learners. The method used here can be applied to WBT content for various subjects.

We now plan to develop content of geography and geology based on the results of the high school teachers' evaluation. Moreover, we will also attempt to integrate our Web content with a Geographical Information System (GIS). This will enable us to provide more effective WBT content by using statistical data from GIS. Furthermore, we will develop more interactive methods for WBT.

ACKNOWLEDGEMENTS

The authors thank Mr. Jun Kobayashi, Mr. Taisuke Suzuki, Mr. Shingo Hinata and Mr. Hideki Wada at Nihon University and Ms. Kaori Suzuki and Mr. Yuichi Yamada at Toyo University for their fruitful discussions and assistance. The authors also express sincere gratitude to the teachers of Waseda University Senior High School, in particular, Mr. Toru Matsuzawa and Mr. Mamoru Takezawa, for their kind cooperation of evaluation of our WBT content

7. REFERENCES

Anada, K., Kobayashi, J., Tsuchida, K., Miyadera, Y., Motohashi, T. and Yaku, T. (2004). A 3-D display system and its data structure for geology education, Proceedings of the 29th Annual Conference of Japanese Society for Information and Systems in Education, 117-118.

Nonaka, I. and Takeuchi, H. (1991). The Knowledge Creating Company: How Japanese Create The Dynamics of Innovation, NY: Oxford University Press.

Ogushi, M. (2003). Knowledge management for education, Proceedings of the 38th Annual Conference of Japan Educational Administration Society.

Training School Managers Works!

Adrie Visscher, Roel Bosker, Martien Branderhorst
1. University of Twente, The Netherlands, 2. University of Groningen, The Netherlands,
3. Ministry of Education, Culture and Science, The Netherlands

Abstract: Although most secondary schools use management information systems (MISs), these systems are primarily used for clerical purposes and are not used to a significant extent to support higher order managerial activities. This situation is less than desirable, as schools are receiving more discretion in developing school policies and MISs could be offering important assistance in this respect. Our research shows the positive effects of a deliberately designed training course on the knowledge of school managers about MISs, on their attitudes towards MISs, and on the skills school managers need in order to use MISs to support decision-making.

Keywords: School management, management information systems, training course.

1. INTRODUCTION AND PROBLEM STATEMENT

Most secondary schools use computer-assisted information systems in their organisational operations, however, their use tends to be heavily concentrated in the area of clerical functions (Fung, et al., 2001; Visscher & Bloemen, 1999). Although this type of assistance is very important, resulting in valuable improvements in efficiency, the information systems (ISs) are not generally applied optimally to the support of higher order managerial work in schools.

The need for informed policy-making has increased in recent years, and it is likely to continue to increase as schools receive more discretion in developing their own school policies (Chubb & Moe, 1990). This rise in school autonomy means that schools must now develop plans in areas in which they formerly executed policies that had been developed at the national level. In order to make optimal use of this room for policy-making, school staff must acquire information upon which plans and decisions can be based. Management information systems (MISs) may provide a basis for more informed decision-making, as various types of computer-produced information can be used in school decision-making for so-called ill-

Please use the following format when citing this chapter:

Visscher, A., Bosker, R. and Branderhorst, M., 2007, in IFIP International Federation for Information Processing, Volume 230, Knowledge Management for Educational Innovation, eds. Tatnall, A., Okamoto, T., Visscher, A., (Boston: Springer), pp. 89–97.

structured problems (e.g. the problem that student achievement levels are too low). Ill-structured problems can have a wide variety of potential causes, and various possible problem-solving strategies of unknown effect exist. Structured problems differ from ill-structured problems in the fact that the former type of problem is characterized by a limited number of variables and solutions, and by known strategies for solving them (e.g. the construction of timetables). Problem diagnosis and the search for solutions are crucial in dealing with ill-structured problems, and these can be supported by the output from ISs in four ways (Visscher, 1996):

- Analysing relationships between variables, e.g. between truancy and student achievement; or between achievement and lesson drop out.
- Analysing patterns over time, e.g. in student intake, staff illness, truancy over several years.
- Answering what-if questions, e.g. how many students will be promoted if the student promotion criteria are raised? How much money will the school receive if the number of students decreases by %? How many teachers will be needed if students are promoted?
- Information system-based policy *evaluation*, e.g. to what extent has the percentage of grade repeaters increased since the adaptation of the promotion criteria? Has truancy increased since the changing of the timetable? What was the effect of extra mathematics lessons on student achievement?

If the policy-making and problem solving performed by school managers improves as a result of the utilization of computing power of MISs, this may improve school effectiveness and is therefore something worth striving for.

An analysis of the research on the introduction of computer-assisted ISs into schools shows that the extent of user training significantly influences the degree to which ISs are used. Table 1 shows that 8 factors predict the extent of IS use one or more times in three different studies. Internal and external training are the only two factors predicting direct use (i.e. use by the person him- or herself, as opposed to indirect use by clerical staff retrieving information from an IS for managers) in three studies in Hong Kong, the Netherlands, England. As mentioned above, ISs are used in particular in the clerical context, in other words, training definitely promotes clerical IS use. Whether this also goes for higher order managerial use of school information systems is something in need of further investigation.

Table 1: Predictors of direct system use in three studies (sources: Visscher et al., 1999; Visscher & Bloemen, 1999; Visscher et al., 2003)

	Hong Kong	the Netherlands	England
Start motivation	x		x
Computer experience	x		x
Internal training	x	x	x
External training	x	x	x
IS data quality	x		x
Clarity innovation goals	x		
Clarity innovation means		x	
Length of personal use			x

There is, however, good reason for focussing on training in order to promote the use of ISs for managerial purposes, as managerial IS use presupposes various skills. Any training course should bring managerial staff to a level at which they are able to decide which information is needed for decision-making, and to interpret the data in such a way that it can be used for developing, implementing, and evaluating decisions and school policies. Managers must learn, for example, what specific data do or do not say, as they are not experienced in interpreting aggregated computer data. Unfortunately, the content of the few principal training courses that do exist tends to be both too technical and too theoretical in nature (Visscher & Bloemen, 1999).

Given the aforementioned state of affairs, the research conducted in this study was guided by the central question as to whether or not a deliberately designed training course can improve the utilisation of MISs by secondary school principals for purposes of managerial support.

2. CHARACTERISTICS OF THE TRAINING COURSE

An important step towards answering the research question involves determining what exactly school managers should be trained for, and how this should be done. This question has been answered by analysing the scientific literature on: a) the implementation of information and communication technologies in education, especially the introduction of MISs; b) strategies for adult training. The literature review indicates (for more details see Bosker, Branderhorst & Visscher, 2006) that the training course should have the following characteristics:

- Match the nature of the training course with the expertise and skills of participants, starting with problems they face in their professional practise.
- Determine the needs of individual participants at the start, and strive to ensure that participants experience maximal success as quickly as possible.
- Teach participants how to determine the kinds of information they need, as well as how they can select, retrieve, interpret, and use it in school policy-making. To this end, make use of various instructional strategies, including active learning, self-study, and group assignments.
- Promote the transfer of what has been learned to professional practice by offering on-the-job support and involving colleagues in the training course.

3. DESIGN OF THE STUDY

The central research question was answered by investigating whether principals who had followed the designed training course know more about the way the MIS can be used to support decision making, whether they use it more for developing and evaluating decisions, and whether they are more positive about the possibilities offered by the MIS. The effect of the training course was evaluated using the Solomon four-group design:

R	O1	X O2	O3	group 1
R	O1	O2	O3	group 2
R		X O2	O3	group 3
R		O2		group 4

R = the four groups are composed by using randomization procedures
X = experimental treatment
O1 = pre-test
O2 = post-test two weeks after the treatment
O3 = retention test three months after the treatment

In the Solomon four-group design, the classical pre-test post-test control-group design is augmented by two extra groups in order to control for the sensitising effect of the pre-test and for any interaction between pre-test and treatment (Philips, 1976; Swanborn, 1999). By adding the retention test, it was possible to determine how and to what extent the effect of the treatment was maintained over a period of three months. In practice, all groups were trained in the end. The school leaders that made up group 2, for instance, were tested at O1 and O2, after which they became part of group 1, for whom O2 was the pre-test. For that reason there were four time points: at T1 a pre-pre-test, at T2 a pre-test, at T3 a post-test, and at T4 retention test.

Currently, three MISs are in use in the majority of Dutch secondary schools. At the time of the study, however, only one MIS, called School+ Web, could generate information to support the management activities of principals (this is another indication of the clerical focus in the support given by ISs in schools) and was therefore chosen as the MIS for which trainees would be trained in the training course.

Data was collected through a written questionnaire. Because of the lack of prior research on the principals' use of MISs for decision-making tasks, it was necessary to develop a questionnaire. This questionnaire was meant to distinguish between three elements of the dependent variable 'the degree of MIS use': knowledge of the MIS, skills in MIS use, and the MIS attitude (factor analysis supported the existence of these three factors, which explained a total of 52% of the variance).

In addition to the three scales (reliability varied between 0.87 and 0.96), the questionnaire included variables that were expected to be related to the dependent variables (for example, age, experience with computers, school size). Principals were also interviewed in order to gather more information about the transfer of skills and knowledge acquired as well as to examine the extent to which these managers generate management information and use it for the policy process.

Managers of schools using School+ Web were informed of the existence of the training course. Those who wanted to follow the training course were assigned randomly to one of the four groups of the Solomon design. The groups can therefore be assumed to be comparable.

With regard to obtaining a sample of sufficient size and scope, data collection took place in the period between autumn 2002 - spring 2004. A total of sixty-one respondents participated in the study (response rate 95%); the number of respondents per Solomon group is: group 1= 12; group 2 = 13; group 3 = 18; group 4 = 18.

The groups of respondents consist mainly of men of an average age of 49. The average size of the schools involved is 1500 pupils. The respondents' use of the computer is very intensive, but that does not apply to the use of management information. Furthermore, the respondents' knowledge and skills in the field of School+ Web are limited. The four Solomon groups appear to be strongly comparable on the context variables.

Analysis of variance and t-tests (level of significance was set at 0.10 one-sided) were used to test the significance of pre-existing differences between various groups. Multilevel models (two levels: moment of measurement, and trainee) were applied to test whether the experimental treatment was related to differences in the growth trajectories of the trainees. (Snijders & Bosker, 1999).

4. RESULTS

The results presented here offer an answer to the question as to whether a deliberately designed training course can improve the utilisation of MISs for managerial support. In more operational terms, data analysis focused on finding out whether the training course produces differences in the knowledge about, attitudes towards, and skills in the use of the MIS between the experimental group and the control group.

The general hypothesis is that, as a result of the training course, the members of the experimental group are more knowledgeable about which information can be retrieved from School+ Web, have a more positive attitude towards the use of the MIS, and have better skills in using MIS than the members of the control group.

4.1 Knowledge

The difference between the experimental and control groups with regard to the knowledge about the management information that the MIS can retrieve was statistically significant, not only at the end of the training (t =

3.50; p<0.001)[1,2] but also a few months later (t = 4.10; p<0.001). These differences remained when controlling for differences in knowledge between participants at the start of the training. The analyses further revealed no significant interactions between treatment and pre-test (t = -1.47; p = 0.144).

The effect size[3], the power of a test after the experiment has been performed (T3 vs T2), is 0.80. For the retention test (T4 vs T2)), the effect size is 0.83. According to Cohen (1988), both effects are large[4].

4.2 Attitude

The effect of the training (T3 vs T2) on the attitude towards MIS use appeared to be significant (t=2.03; p=0.044), with an effect size of 0.44. After a few months, however, the effect (T4 vs T2) had disappeared (t=0.50; p=0.62). In other words, at that time, the attitudes of the members of the control group were the same as those of the experimental group. No significant interactions were observed between treatment and pre-test (t=-1.20; p=0.22).

4.3 MIS use

The analyses of the data concerning the use of the information system for decision support revealed that the mean score of the experimental group is significantly higher than that of the control group, taking scores on the pre-test into account (t=2.82; p=0.006), with a medium effect size of 0.52. After a few months (T4 vs T2), this difference was smaller, but still significant (t=2.00; p=0.05), with an effect size of 0.33. There were no significant interactions between treatment and pre-test (t=-0.48; p=0.63).

4.4 Covariates

The questionnaire included several contextual variables that were expected to be important for the use of ISs for managerial activities. The analyses, however, revealed no differences between the experimental and control groups for these variables, with the exception of the variable 'school size'. Because this difference could have influenced the results of the analyses, the analyses were conducted once more with school size as a covariate. School size was found to have a significant effect on the use of

1 In the empty model, the mean and variance are estimated. In the model, T2 is taken as the contrast for the time points.

2 For studying the significance of effects, the t-ratio has been computed with 50 degrees of freedom (there wère 111 degrees of freedom for the interaction effects).

3 The effect size has been computed by dividing the regression coefficient by the square root of the total variance, which is the residual pooled standard deviation.

4 Cohen defines an effect size of 0.8 as large; an effect size of 0.5 as average, and an effect size of 0.2 as small.

the MIS[5]: the treatment effect for principals in smaller schools was greater than for principals in larger schools.

4.5 Interviews

The main conclusion of the interviews is that, after completing the training, principals expected to be able to use information from School+ Web for decision support, but that all users experienced problems with the information system. Because of technical problems, the MIS failed to meet expectations. The attitudes of the users concerning the possibilities of using School+ Web for developing, implementing and evaluating decisions and school policies had therefore become less positive.

5. CONCLUSION

The results of this study reveal that members of the experimental group are more knowledgeable about which management information the MIS can retrieve, are more positive about the use of MIS, and are more skilled in using the MIS for decision support. However, some months after having completed the training, the attitude of the manager with respect to the use of management information is less positive than a few weeks after the training.

Whether or not the research results may have been caused by the pre-test, or by an interaction between the pre-test and the intervention, was also examined. Although the research results show that the training course has a little more effect if a pre-test has been taken, this effect does not prove to be significant.

The effect of the relevant context variables on the difference between the groups was also analysed and found to be only marginal. It appears that the training effect is somewhat larger when the school is smaller. An explanation for this finding may be that quantitative management information receives less attention in a small school compared with a large school because school managers in small schools have a good overview of how the school functions without that information. As a result, the training course may have more of an effect on the use of management information in small schools because of the prior use was more limited there. However, including the 'school size' variable in the multilevel model does not lead to other outcomes when testing the hypotheses.

In this study, a training course was developed, based on a literature review and in consultation with school managers, with the expectation that it would produce the desired effects. This training course took into account the

[5] The variable 'school size' has been centered around the general mean to facilitate the interpretation of findings.

individual needs of the participants and was designed to fit with the actual work practice of school managers. Unfortunately, the planned training follow-up activities, such as organizing a follow-up day, did not take place. Such activities are, however, important for the transfer of training to work practice. Furthermore, some participants found the differences between the training participants too large in terms of their knowledge of School+ Web, and the degree to which school policies were developed within their schools (before they were trained). It would be interesting to optimize the training course in order to address these shortcomings and to examine whether the training effect would then be even greater.

Now that we know that training courses with specific characteristics have very positive effects on the use of MISs in schools in terms of knowledge, attitude, and skills, the next step should be to conduct more longitudinal research into the question concerning whether a more intensive use of MISs leads to better decisions, and, eventually, to more effective schools.

To which target groups can the research results be generalized? The 'objective population' includes managers of secondary schools who work with a computer-assisted information system and who have an interest in the utilization of management information. However, the experiment was carried out with the information system School+ Web, and the sample was therefore limited to Dutch managers with an interest in management information working in schools that use School+ Web. Theoretically, the generalizability of the results is limited. However, on the basis of interviews with a number of school managers who use a different MIS than School+ Web and who did not participated in the training course, it seems that the use of management information from other information systems is very limited, but that, at the same time, the interest in the subject is growing. With the training course central to this study, it might be possible to promote the use of other MISs. Most schools are still in the implementation stage and, as a result, do not prioritize its use for management information. When they have fully implemented a MIS, it will be interesting to evaluate the impact of a training course like the one we have studied in this project.

Principals make many decisions under conditions of uncertainty. They are burdened with information and have little time to process or reflect on the information. Full rational behaviour, in terms of choosing the best way to achieve explicit goals after having processed relevant information, is rare (Riehl et al., 1992). Principals also do not frequently evaluate the performance of their schools in terms of trends and results. It is therefore not surprising that MISs are not yet used much for higher order decision-making. The results of this research project provide empirical evidence for the assertion that carefully designed training courses can stimulate principals to use MISs for decision support in schools. Hopefully, our findings will encourage and support others responsible for the design and implementation of MISs to improve the quality of management activities within schools.

6. REFERENCES

Akker, J. van den, Keursten, P. and Plomp, T. (1992). The integration of computer use in education. *International Journal of Educational Research, 17*, 65-76.

Baldwin, T.T. and Ford, J.K. (1988). Transfer of training: a review and directions for future research. *Personnel Psychology, 41*, 63-105.

Bosker, R.J., Branderhorst, E.M. and Visscher, A.J. (submitted for publication). Improving the utilization of management information systems in secondary schools. *Journal of Educational Research.*

Chubb, J.E. and Moe, T.M. (1990). *Politics, Markets and American Schools*. Brooking Institute, Washington, DC.

Cohen, J. (1988). *Statistical power analysis for the behavioral sciences*. Hillsdale: Lawrence Erlbaum Associates.

Fung, A., Visscher, A., Smith, D. and Wild, P. (2002). Comparative evaluation of the implementation of computerised school management systems in Hong Kong, The Netherlands and England. In D. Watson and J. Andersen (Eds.), *Networking the learner, Computers in Education* (pp. 591-600), Boston/Dordrecht/London, Kluwer Academic Publishers.

Phillips, B.S. (1976). *Social research; strategy and tactics*. New York: Macmillan Publishing.

Riehl, C., Pallas, G. and Natriello, G. (1992). *More responsive high schools student information and problem-solving*. Paper presented at the Annual Meeting of the American Educational Research Association, San Francisco.

Snijders, T.A.B. and Bosker, R.J. (1999). *Multilevel analysis: an introduction to basic and advanced multilevel modeling*. London: Sage.

Swanborn, P.G. (1999). *Evalueren. Het ontwerpen, begeleiden en evalueren van interventies: een methodische basis voor evaluatie-onderzoek. [Evaluation, Designing, supervising and evaluating interventions: a methodological base for evaluation research.]* Amsterdam: Boom.

Visscher, A.J. (1996). Information technology in educational management as an emerging discipline. *International Journal of Educational Research, 25*(4), 291-296.

Visscher, A.J., & Bloemen, P.P.M. (1999). Evaluation and use of computer-assisted management information systems Dutch schools. *Journal of Research on Computing in Education, 32*(1), 172-188.

Visscher, A.J., Fung, A. and Wild, P. (1999). The evaluation of the large scale implementation of a computer-assisted management information system in Hong Kong schools. *Studies in Educational Evaluation, 25*, 11-31.

Visscher, A., Wild, P., Smith, D. and Newton, L. (2003). Evaluation of the implementation, use and effects of a computerised management information system in English secondary schools. *British Journal of Educational Technology, 34*(3), 357-366.

REFERENCES

Technology Enhancing Learning

Limited data handling facilities limit educational management potential

Don Passey
Department of Educational Research, Lancaster University, UK

Abstract: This paper considers the role of management information systems (MISs) in supporting practices that can lead to enhanced achievement. In England, MISs have not been provided centrally by government or government departments, but have been purchased by schools. MISs have offered resources focused largely for use by managers rather than teachers. That formative assessment can be used by teachers to enhance attainment has been well studied, and a clear link has been recognised. National agencies have, since about 2000, promoted concepts of assessment for learning in educational practice across schools. MISs can provide a key means for schools and teachers to handle, review and monitor formative assessment data. Although some schools use MISs for this purpose, most schools recognise limitations with the system they have, and studies increasingly identify issues and specific limitations. A pilot project shows how innovation can address issues. However, it is clear that there is need for further innovation and development. National policies will need to consider the entire range of challenges, if teachers are to use MISs to support and enhance learning achievement effectively.

Keywords: Formative assessment; management information systems; innovation; learning attainment

1. INTRODUCTION

This paper is concerned with ways in which management information systems (MISs) can support educational practices (at a classroom or pupil level) that can lead to enhanced achievement. In England, MISs have not been provided centrally by government or by government departments, but supplied by a number of companies and purchased by schools. These commercial systems have offered resources focused largely for use by managers rather than teachers.

Formative assessment can be used by teachers to enhance attainment. This fact has been well studied, and a clear link has been recognised. Black

Please use the following format when citing this chapter:

Passey, D., Madsen, P., 2007, in IFIP International Federation for Information Processing, Volume 230, Knowledge Management for Educational Innovation, eds. Tatnall, A., Okamoto, T., Visscher, A., (Boston: Springer), pp. 99–106.

and Wiliam (1998), in reviewing the research literature, looked at some 30 studies, which used experimental and control groups, pre-and post-tests, and provided numerical data about learning gains. They found firm evidence that formative assessment was an essential component of classroom work, and that effective practice and use could raise standards of achievement (with gain sizes in the order of 0.4 to 0.7 of an attainment level obtained when formative assessment practices were used). The Department for Education and Skills (DfES) in England introduced a focus on 'Assessment for Learning', and in 2002, the Key Stage 3 initiative (for pupils aged 11 to 14 years) stressed the importance of taking an 'Assessment for Learning' approach. A part of that emphasis led to the publication of a booklet focusing specifically on data management use (Releasing Potential, Raising Attainment: Managing Data in Secondary Schools, DfES, 2002).

The Qualifications and Curriculum Authority (QCA) outlines on its web-site resources (2005), the importance of taking an 'Assessment for Learning' approach. It states that: "Assessment for learning is the process of using classroom assessment to improve learning, whereas assessment of learning is the measurement of what pupils can do. In assessment for learning: teachers share learning targets with pupils; pupils know and recognise the standards for which they should aim; there is feedback that leads pupils to identify what they should do next in order to improve; …". It goes further by stating that: "Assessment for learning is one of the most powerful ways of improving learning and raising standards. Actively involving all pupils in their own learning, providing opportunities for pupils to assess themselves and understand how they are learning and progressing, can boost motivation and confidence".

MISs have been a key means to help schools manage formative assessment data. A recent research survey that looked at school uses of data for teaching and learning (Kirkup, Sizmur, Sturman and Lewis, 2005) found that: "the impact of data on teaching and learning operates at two levels: directly by means of interventions targeted at individual pupils; and indirectly by means of whole-school approaches". They went on to say that: "Commonly reported uses for data in all schools were: to track pupil progress; to set targets; to identify underachieving pupils for further support; to inform teaching and learning and strategic planning. … At the classroom or pupil level, effective use of data enabled schools to: highlight specific weaknesses for individual pupils; identify weaknesses in topics for the class as a whole; inform accurate curricular targets for individual pupils; provide evidence to support decisions as to where to focus resources and teaching".

2. ISSUES: THE REALITY

In 2002, research undertaken by the author into MIS practices in a key range of schools across a number of local education authorities (LEAs), indicated that although the potential support that MISs could offer schools in terms of enhancing formative assessment practices was high, the reality was that the information technology (IT) systems in place were not functioning in ways to support teachers and schools (Somekh et al., 2002a). The report stated that: " What is needed is a fine level analysis of the issues and the approaches that could be adopted when designing the MIS. ... Currently, lack of sophistication of analysis is leading to a blurring of needs, and is limiting possibilities". Overall, functionality to support teacher curriculum needs was not fundamentally central to the systems observed, data of all forms was held in a single system, differential access was not adequately provided, and, hence, there was no integration of approach based on specific user needs. The report indicated issues that schools faced: "Currently there is a major focus upon data gathering, data input, data records, and data output (rather than upon data transfer and data analysis). The systems often do not allow ease of transfer of data, and analysis of data is often limited to large numerical tables that teachers find difficult to handle and to interpret. While the Pupil Level Annual School Census (PLASC) has enabled data to be recorded in electronic form, schools often experience difficulty in transferring data from the MIS to PLASC. Their understanding of the value of this for their school is often not clear despite the investment of time and effort in carrying out manually what could in theory have been undertaken electronically".

The final report on IT practices across the ten key LEAs, all involved in implementing the major national IT National Grid for Learning (NGfL) roll-out (Somekh et al., 2002b), suggested that innovative approaches needed to be taken: "The NGfL Programme had the potential to revolutionise an LEA's ability to access, manage and make use of information about its schools and their pupils. In practice, however, this is an area where the NGfL has been slow to make an impact". At that time some LEAs were moving towards more central systems, but issues were identified with those approaches: "As with other areas of centralisation, however, the impact on individual teachers in schools is not always immediately beneficial. One school ICT co-ordinator complained that her own MS Excel-based record system was more efficient for tracking pupils' progress and performance than the new system, which the LEA had recently provided".

The recent research survey (Kirkup, Sizmur, Sturman and Lewis, 2005), shows that the situation is in practice similar to that found 3 years earlier. The report highlighted practice associated with effective uses of MIS. For example, it stated that: "Schools reported that effective use of data resulted from meaningful dialogue between staff, and was supported by user-friendly systems". The report indicated that many schools were developing their own methods to account for the limitations they experienced: "Rather than closed data analysis packages, school-devised systems and Excel spreadsheets were the most popular data management tools because they tracked individual pupils and allowed schools the flexibility to input internally generated data such as interim assessments and targets; i.e. such tools were easier to customise to the school and its particular needs and circumstances". Even systems provided centrally by the DfES were not found easy to use by all schools: "Users of the Pupil Achievement Tracker software (PAT) provided by the DfES generally made positive comments about the visual presentation of data and the ability to compare groups of pupils. However, many questionnaire respondents and focus group participants found PAT very difficult to use and were confused as to how to input data". The report concluded that: "All schools wanted data management systems that: are easy to use; produce outcomes that are easy to interpret; allow flexibility of input; have compatible school management and assessment components; offer comprehensive training and support; are accessible to staff; encourage engagement and ownership".

Becta, in a recent review of the state of MISs in England (2005), echoed many issues raised in previous studies. They focused particularly on commercial and management aspects that might enable MISs to be developed further. Becta stated that: "... we estimate that the total cost of providing and supporting MIS systems in schools in England is at least £180 million annually, and could be much higher. We confirm that there are considerable impediments to maximising the potential value for money flowing from that expenditure. Those impediments span all aspects of the current arrangements including the contractual landscape, the technical environment, the support arrangements and the statutory returns process". On the issue of interoperability, Becta reported that: "We find that interoperability arrangements are effectively dependent on the dominant supplier, which sets the detailed technical, financial and legal framework within which interoperability takes place. We consider that, if unchecked, such arrangements for interoperability have the potential to impede competition and choice not only in the provision of MIS solutions but also in the market for Virtual Learning Environments (VLEs) and Managed Learning Environments (MLEs), and hinder the effective delivery of wider policy objectives in relation to personal learning spaces".

3. LIMITING FACTORS: THE CHALLENGES

In 2001, the Standards and Effectiveness Unit of the DfES set up a project called the Schools facing Extremely Challenging Circumstances project (SfECC), the aim being to look at ways that this group of eight schools could work together, with an appropriate support structure, on an improvement agenda. Because the schools were similar in particular background contexts, but widely spread geographically, this presented challenges of how schools could share practice and ideas. The author undertook an in-depth review of how these eight schools were using MIS, as a part of their support for pupils. The report to the DfES (Passey, 2002), highlighted a number of specific issues: schools did not use the same data input devices for management data; they did not have a master data file for holding input data, and for accessing data; the data needs of schools, the

DfES, the QCA and Ofsted were not the same; data transfer was not easily accomplished, and was very time consuming; forms of data output from data management systems were not constructed for teachers to use easily for analytical purposes; data files contained no links to pupil work; and analysis of records of pupil attainments were not linked to a means to review how lesson delivery was undertaken.

While the schools were reasonably well equipped with IT, their data management systems and their uses were more limited. The schools generally were not at the stage of being able to use information provided from curriculum data management systems: there was no central means to collect and send data; analyses and presentations were not aimed at the classroom teacher or specific curriculum users, or in agreed forms to share with others; curriculum analyses and presentations provided by LEA personnel were not easily accessible; a large amount of time was spent by teachers, support staff and senior managers copying, pasting or recreating data that appeared in other places; the systems in use did not analyse or present information in ways to share or link to practice.

Based on the issues identified, a conceptual framework for IT development was constructed. Illustrated in Figure 1, this shows a sequence of pedagogic processes (that move initially from left to right), each linked implicitly with data management and data handling needs. Evaluation processes, through data review, link back to support subsequent lesson preparation.

Preparation	Teaching	Recording	Review	Evaluation
A teacher reviews data about the class, and uses a laptop to prepare lessons	Resources (including web-based resources) are shown via an interactive whiteboard	Classroom practice can be recorded on video	Classroom practice can be reviewed	Resources can be amended Pedagogic emphasis can be reviewed
	Activities can be undertaken on computers by individuals or groups of pupils	Systems allow outcomes or marks to be recorded Pupil work can be captured and linked to marks	Pupil outcomes can be reviewed	

Figure 1: A framework to develop an integrated IT system to support assessment for learning practices

4. WAYS FORWARD: THE INNOVATION

A pilot project to address data handling limitations for this group of schools emerged in this context. An important aim was to produce a system that would enable a sharing across the eight schools, with data being handled in consistent ways to help inform and share practice. The initial method used to trial this idea was a linked spreadsheet system. However, this system did not address data entry and data flow issues that schools faced. A system now being developed and piloted, called Supporting Teachers in Assessing, Reporting and Tracking (START), addresses these needs far more.

START was, and is being, developed as a data management facility that will act in the form of an umbrella; it is designed to work with existing data management systems and data management elements, to integrate multiple facets of data management that exist in different places. It is designed to be easy to use, aimed at the classroom teacher specifically, but also the form tutor, head of department, and school managers. A key element in the development has been liaison with teachers and head teachers in schools, and key officers in LEAs. The system provides facilities and features suggested by teachers and managers to support their needs. Overall, the concept of START is based on the proposition that IT can help to support integrated systems far more effectively than has previously been possible.

The START resource provides access to a range of Key Stage 3 and 4 analyses and presentations that offer views of class, form, subject and year group data. The facility takes the user through a series of logical steps, so that data uses are considered in a coherent way: looking at background results, of previous Key Stage attainments, to compare results across subjects; choosing estimated likely outcomes, from a range available, including those in PAT, and those from the Fischer Family Trust; setting targets for each pupil in each subject, by selecting appropriate estimated likely outcomes, or by making an informed decision; reviewing a target summary, to see which targets have been set, and whether they are similar across subjects; checking target history, to see the stage of the process reached, and whether targets match estimated likely outcomes; entering teacher subject assessments on an agreed number of occasions across the year; reviewing teacher assessments, including those for behaviour, attendance, effort, and homework; monitoring results, and seeing how attainment is matching a progression towards targets; looking at added value, calculated on the basis of actual results compared to target levels set; looking at analyses that could inform classroom practice, such as analyses of pupil learning approaches arising from the NFER Cognitive Abilities Tests.

START has been reported by teachers and managers involved in the trial as being easy to use (Figure 2 shows an exemplar page). Regular discussions with teachers and managers involved, and an independent review conducted by Becta, have highlighted ease of use. START is web-based and on-line (to offer ease of access), is dynamic (offering up-to-date subject views), allows users to enter data (and to export data into MS Excel systems), provides analyses and presentations (including those recommended by the DfES), works to a background data management calendar, provides alerts and reminders (when data management functions have not been fulfilled), takes the user through a logical sequence of data management events, allows target setting to be done on-line (in a range of ways), enables users to ask questions about the data and information that is seen, when it is seen, and enables users to be responsive, to suggest ideas to those developing the system.

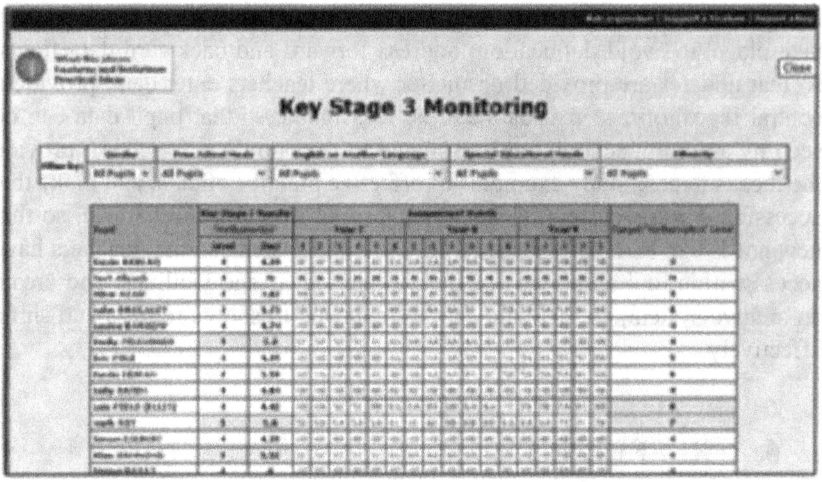

Key Stage 3 Monitoring

Figure 2: Exemplar page that teachers have reported as supporting their classroom practice

The purpose of the START pilot was largely to explore whether issues identified by teachers and managers could be addressed. Through the work of the pilot it is possible to see that many issues can be addressed. Since the outset of the pilot, there have been other significant developments: in forms of data presentation and analysis offered by some MIS and e-learning resource providers; forms of estimated likely outcomes from the Fischer Family Trust; and learning to learn analyses developed by individual schools. Further issues still remain, and moving forward will only be easily possible if the climate is concerned with data flow and transfer, rather than data creation, analysis or presentation.

5. CONCLUSIONS

What has been achieved by a pilot that has sought to address key concerns with the linkage between data management systems and informing teachers at an assessment for learning level is: the creation of an on-line system that teachers find easy to use; analyses and presentations in formats that non-statistical specialists can understand and use; analyses and presentations that can inform teacher practice in classrooms; the import of data from a range of sources, so that access occurs through a single facility; a system that works to a background calendar, and takes teachers through a logical sequence of data management events; data entry linked to presentation so that views are up-to-date; facilities that allow teachers to ask questions via email about the data, as they see it; and a system where teachers can make further suggestions and recommendations.

What has yet to be achieved, and should be a firm focus in the future for any further pilot or national development work is a need to: address data flow, so that data flows from one system to another, to update as changes take place, to avoid duplication; address forward and backward data flows, so that updates are provided no matter where teachers enter data; provide a central repository, so that data can be reassigned, so that pupil data can be seen by a current school, and reassigned to the pupil's next school; provide the means to case study examples of effective practice of data, and make this accessible via the system itself; create a rapid response mechanism, so that new needs can be incorporated rapidly into the system, so that teachers have access within limited periods of time to new ideas and facilities; and create an archive system, so that data can be reviewed to show progress and shifts effectively.

6. REFERENCES

Becta (2005) School Management Information Systems and Value for Money: A review with recommendations for addressing the sub-optimal features of the current arrangements. Becta: Coventry

Black, P. and Wiliam, D. (1998) Inside the black box: Raising standards through classroom assessment. Phi Delta Kappan, 80 (2), 139-148. Retrieved July 30, 2005, from http://www.pdkintl.org/kappan/kbla9810.htm

Department for Education and Skills (2002) Releasing Potential, Raising Attainment: Managing Data in Secondary Schools. DfES: London

Kirkup, C., Sizmur, J., Sturman, L. and Lewis, K. (2005) Research Report No 671: Schools' Use of Data in Teaching and Learning. Department for Education and Skills: Nottingham

Passey, D. (2002) Schools in Exceptionally Challenging Circumstances: ICT Audit. Lancaster University: Lancaster

Qualifications and Curriculum Authority (2005) Assessment for learning. Retrieved 2 October, 2005, from http://www.qca.org.uk/10009.html

Somekh, B., Woodrow, D., Barnes, B., Triggs, P., Sutherland, R., Passey, D., Holt, H., Harrison, C., Fisher, T., Flett A. and Joyes G. (2002a) ICT in Schools Research and Evaluation Series – No.10: NGfL Pathfinders Second Report on the roll-out of the NGfL Programme in ten Pathfinder LEAs. DfES and Becta: London

Somekh, B., Woodrow, D., Barnes, B., Triggs, P., Sutherland, R., Passey, D., Holt, H., Harrison, C., Fisher, T., Flett A. and Joyes G. (2002b) ICT in Schools Research and Evaluation Series – No.11. NGfL Pathfinders Final Report on the roll-out of the NGfL Programme in ten Pathfinder LEAs. DfES and Becta: London

Individual Learning Pattern Related to Intention
with a Biological Model of Knowledge Construction

Toshie Ninomiya , Wataru Tsukahara, Toshiaki Honda and Toshio Okamoto
[1] *Graduate School of Information Systems, University of Electro-Communications, Japan,* [2] *College of Education, Ibaraki University, Japan*

Abstract: In an e-learning system, it is difficult to grasp the learner's condition for learning. In the face-to-face learning environment, we tend to notice if there is something unusual about learners and help them out of difficulties. In addition, we often cannot keep learners from dropping out of the e-learning course. This suggests a requirement for a research study specially focused on how to predict the learner's condition with the learning log data in the e-learning system. Therefore, we drew up a learning model with biological knowledge from the latest molecular biology and brain science. In this paper, results from our learning model were verified by comparison with real learning log data in an e-learning system. Our model suggests that there is a learning type with high intention who prefers to learn in a short term in order to construct his/her knowledge. Then we examined the data related to intention, and it correlated closely with the learning period of one exercise.

Keywords: Knowledge construction, learning pattern, biological model, dopamine.

1. INTRODUCTION

Originally, humans are by nature animals capable of learning in social life. The primordial form of learning is "learning by doing." However, the concept of literacy appeared, thus leading to the production of modern schools at which students participated in unique learning forms. That is, schools were originally used as a "virtual" space for learning. Gradually, school life became an important part of social life. Knowledge acquired in school has been thought of as a ticket to success, but as it is a space for learning, unfortunately various contradictions to the school form have been showing up with the development of information technology. This is due to information technology having the capacity to allow humans to have real and up-to-date knowledge.

Taking into account the historical background, the potential of the e-learning system for natural human learning should be studied. The ability to

Please use the following format when citing this chapter:

Ninomiya, T., Tsukahara, W., Honda, T. and Okamoto, T., 2007, in IFIP International Federation for Information Processing, Volume 230, Knowledge Management for Educational Innovation, eds. Tatnall, A., Okamoto, T., Visscher, A., (Boston: Springer), pp. 107–114.

learn anytime and anywhere has been improved by information technology and researchers begin to discuss the learning theory with constructivism in a new educational way by e-learning. For example, as an environment for situated learning, Grabinger and Dunlop (1995) constructed "Rich Environment for Active Learning (REALs)," and Nunes and McPherson (2002) built "Continuing Professional Development Education (CPDE)" as a new continuous vocational education.

Developing those concepts, the learning grid is a current topic. For instance, the European Learning GRID Infrastructure (ELeGI) project background of learning theory by Vygotsky, Lave and Wenger (ELeGI, 2005). In the ELeGI project, a new learning environment must be discussed. This is the most important topic in making a new learning model. Looking back on the primordial form of learning, a desire of learning comes naturally and humans can learn by themselves with information technology according to the concept of the learning grid.

Here, we would like to think about human desire for learning as coming from nature. We are able to feel the learner's enthusiasm in the face-to-face learning environment of, for example a classroom in school, and thereby support them in an appropriate manner to construct their knowledge. On the other hand, it is difficult to grasp the learner's condition in an e-learning system. Next, we focused on how to predict the learner's condition with the learning log data in an e-learning system.

First, we generated a biological model of knowledge constructions with three factors: emotion, memory and intention (Ninomiya et al. 2005). Each factor simulates chemical reactions which are thought to relate strongly to the latest models in molecular biology and brain science. The results from simulating the learning model suggest that there is a typical learning pattern in those with a high intention personality. In addition, a person of this type is likely to prefer to learn in the short term in order to construct his/her knowledge effectively.

Next, an e-learning course, where learners could access anytime and anywhere, was put into practice. Learners could work out their own plan and carry it out as they like. Although exercises and tasks were given at regular intervals, the deadline of all the assignments was three months and a day. From the learner's viewpoint, the freedom of learning is guaranteed and learners could deal with the exercises and tasks whenever they want. This model suits our aims to get learning log data that originates from the natural desire of learning.

Finally, we examined the learning log data in an e-learning course. In our biological model, the learning pattern suggested that a person with high intention starts constructing knowledge. By contrast, a learner with middle intention starts constructing knowledge after three instructional stimuli. Then, we supposed that the high intention person accesses less than two times for one exercise or task, and that the number of accesses to each exercise or task is one index for the strength of a learner's intention. As well as this typical learning pattern, a person with high intention prefers to learn within a short period of time in the biological model. Therefore, as other indices we chose the number of days between first and final access to one exercise or task and the number of days from task-done to deadline, and it was found that the number of accesses is correlated closely with the number of days from task-done to deadline. Considering these results, there is some possibility of predicting the learner's condition with our biological learning model, and we can provide learners with a suitable learning environment by predicting their learner's condition in the future.

2. BIOLOGICAL LEARNING MODEL

2.1 Biological factor and simulation model

Our model of knowledge construction incorporates both the biological mechanisms of the human brain and the environment to which a learner is exposed in the growing and learning period. First, a few important biological mechanisms related to cognition were selected from knowledge gained in pioneering studies. Furthermore, as there is no doubt about the influence of gene–environment interactions in human development (Botto and Knoury 2001), the influence of environment, including education, was integrated into the model.

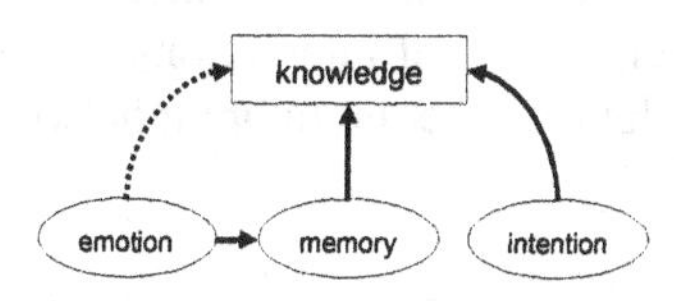

Figure 1. Knowledge model with three factors

We chose three factors: memory, emotion and intention, as there have been sufficiently large numbers of medical case reports (Damasio 1994, 1999) and imaging studies (Bush 2000) to indicate that complicated interactions between these factors affect human cognition. Thus, under the assumption that cognition is influenced by memory, emotion and intention, we made a model of knowledge construction as shown in Figure 1.

To simulate human cognition-related metabolism, several chemical substances that affect emotion, memory, and intention were chosen as shown in Table 1 (Milner 1998, Bliss 1993, Kandel 2001, Poo 2001, Egan 2003, Tang 1999, Dubnau 2003, Gross 2002, Murphy 2001, Gainetdinov 1999, Hariri 2002, Swanson 1998, LaHoste 1996).

Table 1. Chemical substances related to three factors

Factor	Definition	Chemicals
emotion	moods or feelings such as delight, anger, sadness and happiness	serotonin
memory	brain networks that support episodic memory, logical mind, *etc.*	BDNF, NMDA
intention	motives such as wish and desire	dopamine

Figure 2 shows the addition of metabolism to the three-factor knowledge model shown in Figure 1. In Figure 2, ellipses indicate genes and metabolic events such as secretion and absorption of chemical factors are indicted by hexagons. The volume of knowledge, memory, and chemical substances, such as serotonin and dopamine, are shown as rectangles.

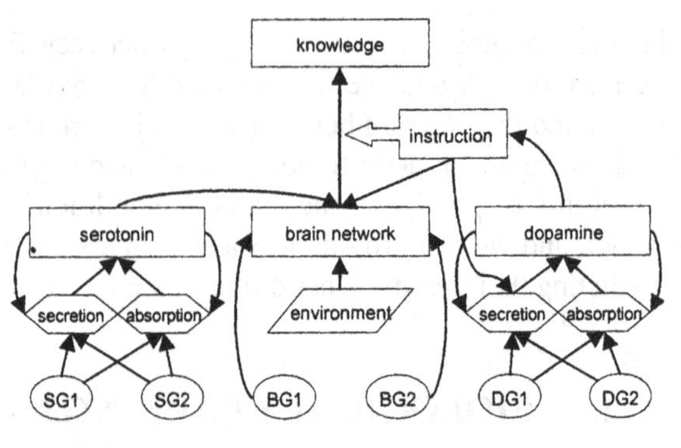

Figure 2. Metabolism model of knowledge construction

Early observations regarding differences in learning ability that emerge from enriched and restricted environments indicated that there are genotype-environment interactions (Cooper 1958). Therefore, the environment control device is set up in brain networks, in addition to those for genes.

2.2 Results of simulation model

In our biological model of knowledge construction, the relation between learning pattern and dopamine metabolic rate is shown in Table 2. Other factors, such as serotonin, brain network, environment, and instruction, have no effect on these learning patterns.

Table 2. Learning pattern and dopamine metabolic rate

Dopamine metabolic rate	Learning pattern
Low (on=0, off=2)	get NOTHING from instructions (knowledge=0)
Intermediate (on=1, off=1)	get knowledge from THIRD instruction
High (on=2, off=0)	get knowledge from BEGINNING of instruction

Figure 3a. Final amount of knowledge (with enriched environment)

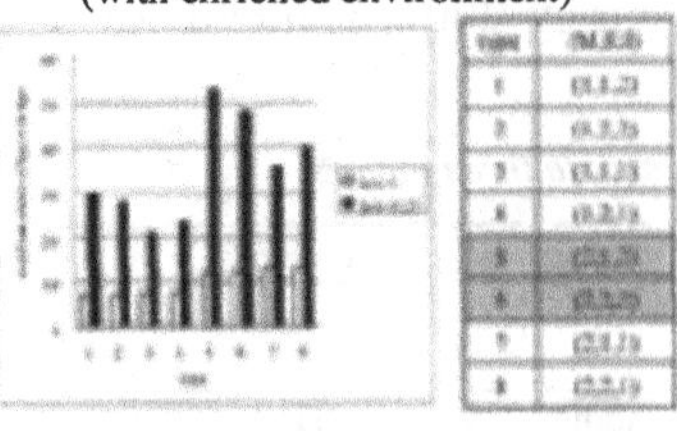

Figure 3b. Final amount of knowledge (with restricted environment)

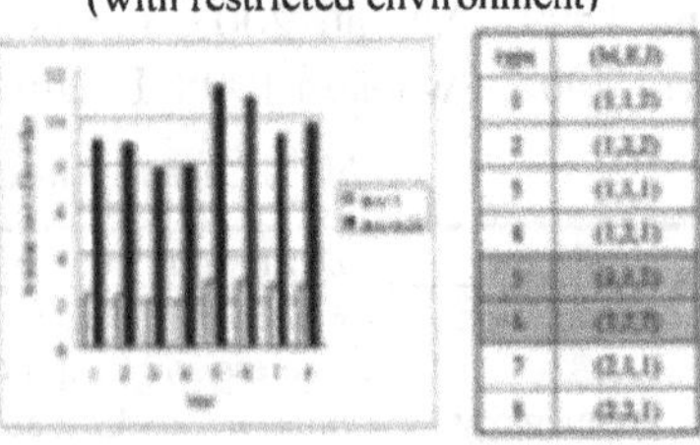

When the interval of instructional stimuli is four times as much as Int.=1, several types benefit from intensive courses. These results are shown in Figure 3a,b.

In particular, types 5 and 6 benefit a great deal from intensive instructional stimuli. These types have a high dopamine metabolic rate as well as an elevated rate of building brain networks. In those with an intermediate metabolic brain network rate, the types with high dopamine metabolic rates also show an increased final amount of knowledge.

3. PRACTICE IN AN E-LEARNING COURSE

An e-learning course of duration three months was put into practice and learners could access the course anytime and anywhere. The outline is shown in Table 3. Thirty-six persons took this course to earn credits for graduation of a postgraduate course. Before starting the course, an orientation and a lecture were given. The learners were allowed to make their own learning plan and to carry out the exercises and tasks as they liked. Although exercises and tasks were given once a week constantly, all deadlines were set a day after three months. This meant that learners had the opportunity to plan their learning schedule and their way of study freely.

Table 3. Outline of the e-learning course

	content
period	2005/05/02-2005/08/09
subject	Artificial Intelligence and Knowledge Management
Topic	9 sub-subjects
exercise	27 tests
Task	9 reports
learner	36 persons (undergraduates, postgraduates and working members of society)

In our biological model, a person with a high dopamine metabolic rate starts constructing knowledge at once. By contrast, a person with a middle dopamine metabolic rate starts constructing knowledge after three instructional stimuli. Moreover, we supposed that a high dopamine metabolic rate person accesses less than two times per exercise or task, and the number of accesses to each exercise or task is construed to be one index of the strength of a learner's intention. The index is expressed by the term "access." As a person with a high dopamine metabolic rate prefers to learn within a short-term in the biological model, the number of days between first and final access with one exercise or task was chosen as the second index. This index is expressed by "period". As well as these two indices, two other indices were examined: the number of days between presenting the tasks and first access and the number of days between final access and deadline. These indices are expressed by "start" and "finish", respectively.

First, we examined the learning log data in the above course between exercises and tasks. Exercises such as short tests were easier than tasks in which learners had to investigate some papers or write a program for a computer. As they were different kind of tasks, it was necessary to check the correlations between these tasks in each index (Table 4). The means of each index for a person was used for calculations.

Table 4. Correlations between different kinds of tasks

	access	period	start	finish
all – test	0.58	0.86	0.99	0.98
all - report	0.89	0.38	0.95	0.93
test - report	0.89	-0.12	0.89	0.86

The important indices "access" and "period" showed low correlations, so correlations between indices were examined for every kind of task (Table 5).

Table 5. Correlations between each index

	All			Test			Report		
	period	start	finish	period	Start	finish	period	start	finish
access	0.24	0.07	-0.41	-0.10	0.13	-0.12	0.61	-0.07	-0.47
period		0.46	-0.39		0.47	-0.33		-0.24	-0.48
start			-0.78			-0.80			-0.63

There are low correlations in Table 5 because the sample size is likely to be small for this investigation. Then the log data was divided into subgroups for each index. For the access index, learners were divided into four sub-groups: 1) less than one access, 2) less than two accesses, 3) less than three accesses and 4) three or more accesses. For the period index, learners were divided into four sub-groups: 1) less than four days, 2) less than eight days, 3) less than twelve days and 4) twelve or more days. For the start index, learners were divided into seven sub-groups: 1) within one week, 2) within two weeks, 3) within three weeks, 4) within four weeks, 5) within five weeks, 6) within six weeks and 7) after six or more weeks. For the finish index, learners were divided into eight sub-groups: 1) within two weeks, 2) within three weeks, 3) within four weeks, 4) within five weeks, 5) within six weeks, 6) within seven weeks, 7) within eight weeks and 8) before eight or more weeks. The correlations between the four indexes are shown in Table 6 after grouping sub-groups.

In Table 6, highest correlations are seen between the access index and the finish index. Therefore, we plotted the relation between these two indexes in Figure 4 below. Regardless of the number of days until the deadline, people with fewer accesses tend to finish their tasks as soon as possible. That is, it is likely that a person with high dopamine metabolic rate prefers to learn within a short term. As even the person with high dopamine metabolic rate accesses more than one with reports, the correlation of report between access and period is the highest as well.

Table 6. Correlations between each index after grouping

	access		
	period	start	finish
all	0.77	0.69	-0.99
test	0.55	0.99	-0.92
report	0.99	-0.25	-0.98
	period		
	access	start	finish
all	0.95	0.51	-0.99
test	0.97	0.87	-0.96
report	0.31	-0.68	-0.82
	start		
	access	period	finish
all	-0.01	0.79	-0.88
test	-0.05	0.76	-0.90
report	-0.15	-0.48	-0.78
	finish		
	access	period	start
all	-0.90	-0.86	-0.85
test	-0.20	-0.67	-0.85
report	-0.88	-0.71	-0.78

Figure 4. Relation between "access" and "finish"

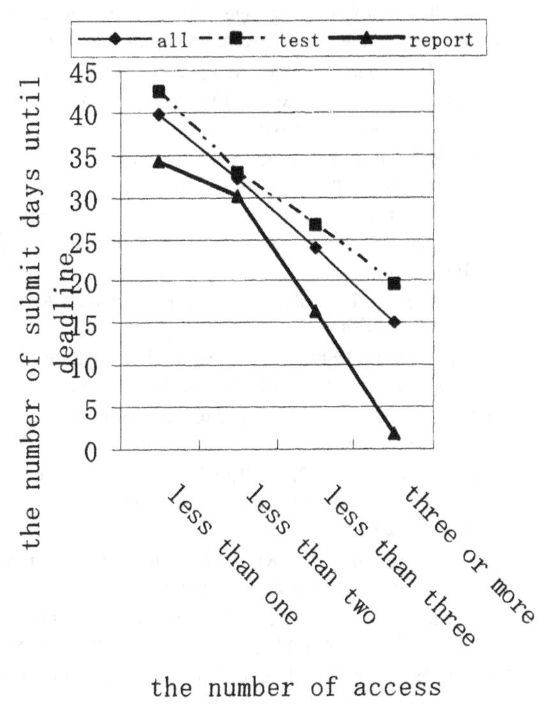

With these results, it was concluded that the number of accesses is correlated closely with the number of days from task-done to deadline. Therefore, there is some possibility of predicting the learner's condition with our biological learning model, and we can provide learners with a suitable learning environment by predicting the learner's condition in future.

4. CONCLUTION

The results of simulations in the biological and social model of knowledge construction suggested that a certain inborn biological function must determine the personal learning pattern, regardless of other functions and environmental influences. Furthermore, intensive training was shown to be effective in high dopamine metabolic rate types of learning. With these simulation results, we examined the data related to intention. It was found that the number of accesses is correlated closely with the number of days from task-done to deadline. Thus, there is a possibility of predicting the learner's condition with our biological learning model.

5. REFERENCES

Bliss, T.V. and Collingridge, G.L., 1993. A synaptic model of memory: long-term potentiation in the hippocampus. *Nature,* Vol.361, pp.31–39.

Botto, L.D. and Khoury, M.J., 2001. Facing the challenge of the gene-environment interaction: The two by for table and beyond. *American Journal of Epidemiology*, Vol.157, pp.1011–1020.

Bush, G. *et al.*, 2000. Cognitive and emotional influences in the anterior cingulated cortex. *Trends in Cognitive Sciences*, Vol.4, pp.215–222.

Cooper, R.M. and Zubek, J.P., 1958. Effects of enriched and restricted early environments on the learning ability of bright and dull rats. *Canadian Journal of Psychology*, Vol.12, pp.159–164.

Damasio, A.R., 1994. *Descartes' Error: Emotion, Reason and the Human Brain.* Putnam Pub Group Published, NY, USA.

Damasio, A.R., 1999. *The Feeling of What Happens: Body and Emotion in the Making of Consciousness.* Harcourt Published, CA, USA.

Dubnau, J., *et al.*, 2003. The staufen/pumilio pathway is involved in Drosophila long-term memory. *Current Biology*, Vol.13, pp.286–296.

Eagan, M.F., *et al.*, 2003. The BDNF val66met polymorphism affects activity-dependent secretion of BDNF and human memory and hippocampal function. *Cell*, Vol.112, pp.257–269.

ELeGI org., 2005/7. GRID Technology for Learning: the "European Learning GRID Infrastructure" project. http://www.elegi.org/.

Gainetdinov, R.R., *et al.* 1999. Role of serotonin in the paradoxical calming effect of psychostimulants on hyperactivity. *Science*, Vol.283. pp.397–402.

Grabinger, S. and Dunlop, J. 1995. Rich environments for active learning. *Association for Learning Technology Journal*, 3(2), pp5-34.

Gross, C., *et al.*, 2002. Serotonin 1A receptor acts during development to establish normal anxiety-like behavior in the adult. *Nature*, Vol.416, pp.396–400.

Kandel, E.R., 2001. The molecular biology or memory storage: a dialogue between genes and synapses. *Science,* Vol.294, pp.1030-1038.

Hariri, A.R., *et al.*, 2002. Serotonin transporter genetic variation and the response of the human amygdala. *Science*, Vol.297, pp.400–403.

LaHoste, G.J. *et al.*, 1996. Dopamine D4 receptor gene polymorphism is associated with attention deficit hyperactivity disorder. *Molecular Psychiatry,* Vol.1, pp.121–124.

Milner, B. *et al.*, 1998. Cognitive neuroscience and the study of memory. *Neuron*, Vol.20, pp.445–468.

Murphy, D.L., *et al.*, 2001. Genetic perspectives on the serotonin transporter. *Brain Research Bulletin,* Vol.56, pp.487–494.

Ninomiya, T. *et al.*, 2005. Social and biological model of cognitive development and learning pattern. *CELDA 2005,* pp.18-25.

Nunes, J.M. and McPherson, M.A. 2002. Pedagogical and implementation models for e-learning continuing professional distance education (CPDE) emerging from action research. *International Journal of Management Education*, 2(3), pp16-25.

Poo, M.M., 2001. Neurotrophins as synaptic modulators. *Nature Reviews Neuroscience,* Vol.2, pp.24–32.

Swanson, J. *et al.*, 1998. Cognitive neuroscience of attention deficit hyperactivity disorder and hyperkinetic disorder. *Current Opinion in Neurobiology,* Vol.8, pp.263–271.

Tang, Y.P.,. *et al.*, 1999. Genetic enhancement of learning and memory in mice. *Nature,* Vol.401, pp.63–69.

Information Technology for Education Management and Open Source Software

Improving Education Management through Open Source

Jacques Bulchand, Javier Osorio & Jorge Rodríguez
Las Palmas de Gran Canaria University, Spain

Abstract: Open Source Software has received lately a great deal of attention, specially due to its lower cost in comparison to Proprietary Software. In the education area, this is quite important due to economical restrictions. Lately, we have seen different Spanish communities embracing the OSS model following different models. This article begins by examining OSS history, as well as its main strengths and weaknesses. It follows examining the possibilities and advantages of OSS in education and presents three possible ways in which the OSS can be introduced in a territory.

Keywords: Open source, education management.

1. INTRODUCTION

There is no doubt that open source software (OSS) is a fashionable topic at present in the technological sector, together with others such as connectivity ubiquity, the improvement of man-machine interaction or future possibilities of Internet.

Before continuing it is important to establish the fact that open source software is also known as free software, but this expression creates certain confusion between two matters, free access and absence of cost. We live in a money-driven world. In such a context, the public's attention is easily drawn towards free things. Due to this there is a general perception that open source software is interesting because of its low or inexistent cost in comparison to owner software, when actually this is not so. In fact, when Richard Stallman, one of the founders of the open source software movement, first approached the subject its economic aspects were not even mentioned.

In this article first of all we will set out the fundamentals of the very concept of open source software, its main advantages and disadvantages and

Please use the following format when citing this chapter:

Bulchand, J., Osorio, J. and Rodriguez, J., 2007, in IFIP International Federation for Information Processing, Volume 230, Knowledge Management for Educational Innovation, eds. Tatnall, A., Okamoto, T., Visscher, A., (Boston: Springer), pp. 115–122.

its present situation in general terms. Having done this, we will look into the possibilities open source software offers in educational environments and how it can be applied in such environments.

We finish by proposing three possible models in which the decisions to implement OSS in a Community can be taken: a top-down model, a citizen pressure group model and a combined model.

2. OPEN SOURCE SOFTWARE

The first matter we must clarify is what open source software is. We will use the definition given by Richard Stallman, considered the founder of OSS. According to this author, OSS must comply with four basic freedoms (Stallman, 1998):

- Freedom 1. To be able to use the program for any purpose, that is, there can be no limitations imposed on receivers when selling or handing over a program to them as regards what they can and can't do with that program.
- Freedom 2. To be able to study how the program works and to adapt it to our own needs, that is, its code must be susceptible of analysis and allow changes, as well as the possibility of creating new programs tailored to suit our needs.
- Freedom 3. To redistribute as many copies as we wish, with or without charge.
- Freedom 4. To improve the program and to give the community access to these improvements so that the whole community benefits from them. Conceptually speaking, this freedom is very similar to the second one. The difference for Stallman derives from the fact that this freedom must also allow offering the community the improvements made.

We must keep in mind that freedoms 2 and 4 require the receiver to have access to the program's source code and, as we previously mentioned, Stallman does not mention whether OSS should be free of cost or not. Therefore, there is a possibility that OSS may be charged for. However, once purchased it can be used in whatever way we choose: introducing modifications, correcting it, distributing it and so on.

2.1 OSS history

To intend to write the history of open source software seems pretentious at the very least. Nevertheless, we will try to briefly outline the historical evolution of OSS.

In the Sixties and Seventies software was considered irrelevant in Information and Communication Technology to the point that manufacturers sold the hardware and gave away the software. But after the anti-monopoly lawsuit against IBM, the courts forced a separation of these components so they started to be billed separately to clients. As from that moment software started having its own value.

Because of this by 1980 most software was protected by intellectual property rights. For security reasons most program owners decided to prevent general access to their source codes and forbade their programmers from talking about them to anyone. They considered this would contribute to keeping the secrets of their creations, just as with any other element subject to intellectual property rights. Obviously, bit by bit these decisions hindered cooperation between programmers.

When in 1986 Richard Stallman, an MIT researcher, was asked to sign one of these non-disclosure agreements he decided to publish the GNU manifesto. By doing so he embarked on a project to create an operating system compatible with UNIX (GNU is not UNIX) but with a difference: it would be totally free and absolutely open to alteration according to user's specific needs (Stallman, 2000).

The project developed and grew stronger. Many programmers enthusiastically took to the idea of creating free software and after several years of hard work they had developed the basics of an operating system: a compiler, a language based editor, an interpreter and tools for networking. In spite of this, the main element was still missing, the very heart of any operating system, known as the kernel. This would be the main contribution that the Linux system would make to the project.

In 1990 Linus Torvalds, a student of the University of Helsinki (Finland) decided to improve an operating system called Minix that exploited the potential of the newly arrived 80386 to the maximum. This is how he created Linux, so well known nowadays, and also how Stallman and his collaborators found the missing element for their operating system, its kernel. By merging both what we now know as GNU/Linux was born.

From that moment new open source software begins to appear until we reach the present situation in which this kind of software is available for almost any application, such as office automation (word processing, presentations, spreadsheets, etc), graphics, statistics, mathematics, design, distance learning, web services and e-mail, amongst others.

2.2 Advantages, disadvantages and risks of OSS

It seems appropriate to briefly cover the main advantages, disadvantages and risks associated to the use of open source software. Amongst the advantages we can state the following:

- Cost (Fink, 2003 and Watson *et al*, 2005). Even though we have already mentioned that freedom of use must not be confused with non-payment, the fact remains that license models contribute to significantly reducing the final costs of open source software as opposed to owner products.
- Independence from seller (Fink, 2003 and Weber, 2004). Obviously, having access to the program's source allows users, if necessary, to change sellers, develop their own versions, etc.

- Control and quality (Fink, 2003 and Watson *et al*, 2005). By having access to the program's sources, we can always know what our software can do or what it includes and examine possible faults, which enable us to carry out quality audits at any moment by means of *a posteriori* verification. As well as this, the fact that so many people take part in the development of each project ensures a higher quality of the final product, rather than when only one person or a reduced group are in charge, even if they are professionals.
- Innovation and development possibilities (Fink, 2003; Raymond, 2001 and Watson *et al*, 2005). Once again, access to source codes will allow personal alterations without depending on third parties. In turn, this encourages innovative capacities that are not linked to any particular supplier.
- Availability of qualified personnel and local promotion of the technological industry (Fink, 2003; Pavlicek, 2000 and Bulchand, 2005). Access to source codes together with the fact that products are widely known promotes quick training of technical experts which, in addition, are generally available. Also, access to source codes supports the development of local industries.

There are other less important advantages of OSS which are, nevertheless, worth mentioning. Firstly, technical support is widely available (Fink, 2003) because products are or can be well known; the option of trying the product before purchasing it; and last, but not least, OSS is associated with an ethical advantage that avoids excessive economies of scale and allows wealth to be proportional to the work put in.

There are, in spite of all these advantages, a few drawbacks to OSS, most of which are closely linked to the very causes of those same advantages. Some of these disadvantages are:

- Availability of applications for the operating system (Fink, 2003). Presently there are not so many applications available for Linux as for other systems such as Windows. However, there are also several OSS products that have been developed for proprietary systems, though in such a case the final result would be a hybrid that would include some of the advantages of OSS but lack others.
- Maturity (Fink, 2003). The fact that there is not one single firm responsible for development means that sometimes we will find on the market products which are under trial, evaluation or similar. Frequently the success or failure of a project is based on such an unsystematic factor as the number of organisations that adopted the project in its early stages. Sometimes early-adopters can find themselves in trouble if a group or community does not grow around the project. Recently we have seen an example of this concerning e-learning platforms. There were two big commercial platforms on the market, WebCT and Blackboard, which incidentally have now merged. Simultaneously, several OSS e-learning projects appeared: Moodle, Ilias, Claroline, Atutor, etc. Whereas choosing a commercial platform at least apparently

offers certain guarantees, deciding on an open one can certainly be risky if a community has not yet formed around it.

- Scalability (Fink, 2003). Sometimes OSS products are suitable in limited production environments for a small number of users, but their possibilities on a large scale are restricted. The reason is that developing goods for a large production environment is complex, as simply the trial stages require installations that are not easily available to open source programmers which many times are freelance or program as a hobby. An example of this is the Oracle database management system as opposed to MySQL.

Finally, OSS is not free of risks, amongst others the following (Golden, 2005):

- Risks associated to license models, because this licences are less known to firms than the traditional ones. As a result, their legal departments do not dominate the situation and feel less comfortable about them.
- Security and quality hazards. OSS products are created, altered and, generally speaking, handled by many people, with greater chances of, for example, having Trojans installed.

3. OSS IN EDUCATION

In general terms, most of the advantages we have mentioned are directly applicable to education. On the other hand, some of the risks and disadvantages are minimised due to the structure of the institutions involved. We will now study them briefly.

As far as advantages are concerned, most of them apply in education. However, we would like to specially draw attention to costs. We must remember that financial pressure is considerable in educational centres of most public systems, both in developed and non-developed countries. Pressure is such that finding a way of computerising schools in a legal way and for a reasonable cost is extremely important. However it would not be fair to disregard the effort made by some manufacturers of owner software, mainly Microsoft, to adjust their prices to the possibilities of educational institutions, mostly because they realise that the germ for future users and technicians is to be found here. As a result, the differences between costs are often not as relevant as one would think.

On the other hand, all advantages concerning development and innovation that are substantial to OSS are also essential in the educational environment, especially regarding technical studies, for which the chance to examine the code, to understand it, alter and improve it becomes a fundamental part of the learning process.

Finally, we must not forget the motivational plus for a student who knows that by analysing the free code of an application the services he can in future offer the market will be much closer to existing needs. His function will go beyond simply pressing the "*Next*" button and he will have access to much more interesting options.

About the disadvantages, none of the ones mentioned above (maturity, availability or scalability) are especially important in educational institutions. Because management mostly involves a reduced scale, here we are not concerned with highly technological applications that require a superior degree of maturity of its processes nor a wide range of options or even a great capacity of scalability.

Neither does one of the risks involved, the one regarding license models, seem especially worrying. The same can not be said for security. The fact that the source code that guarantees, for example, inviolability of marks is an issue that requires a specific debate.

3.1 OSS and teaching and learning

Apart of what has been already mentioned, OSS is about an architecture of participation which can be extended beyond software (O'Reilly, 2006). This software has just been the first area to show the power of self-organizing teams in producing value. In fact, the concept behind Web 2.0 is this architecture, being some examples those of the Wikipedia, Flickr or Amazon, three cases that obtain a good deal of their value from used cooperation.

This means one of the goals teachers have is how to translate the OSS flavour into our systems of teaching and learning, since this will allow students to better collaborate and work together in the way the business environment is going to demand from them in the near future.

3.2 OSS in Spanish secondary education institutions

In the last few years Spanish secondary education institutions have gone through an intense process of adoption of OSS. The Linex Project carried out by the Junta de Extremadura is emblematic and has received ample attention from the media, including the Washington Post (November 3rd, 2002). One of the pillars of this project has been the computerisation of secondary education institutions, as described in the paper "Information Society in Extremadura" (Díaz, 2005).

A similar path has been followed by other Spanish Autonomous Communities, in Valencia under the name Lliurex or in Andalusia known as Guadalinex.

Intuitively the advantages of OSS in secondary education seem obvious, as we have already stated. But many questions arise almost immediately. Why use OSS in secondary schools? Is it just a matter of costs, an ethical issue or is there certain social pressure imposed by Linux organised groups? Are the previously mentioned benefits true in practice?

We believe that the sort of decisions that combine education management and Information and Communication Technology can become interesting guides that will point us towards the future of ITEM in the following years.

4. IMPLEMENTATION MODELS OF OSS

Our approach to the matter suggests three possible models that could lead to implanting OSS in an Autonomous Community.

First of all, a solid political resolve. This would be a top-down model in which the ruler would be firmly convinced of the benefits of OSS, thus starting a process that would gradually involve all social agents. This is the model has been used in Extremadura, which an important financial investment in the past years in OSS technologies as well as in placing the necessary hardware in the classrooms.

The second possibility would originate in the citizenship that, organised in pressure groups and setting a neighbouring community or another country as an example, would pressurise governments to implement this kind of initiative, even though rulers were not especially convinced of its benefits. This model has been tried in various regions in Spain, none of which have been successful up to now.

Lastly, the third would be a combination model in which OSS solutions would initially be put into practise jointly, partly due to a clear interest of the political class but also because of significant pressure exercised by citizens. This is the model used in certain regions that have followed Extremadura, in which citizens have drawn the attention of politicians towards the success in Extremadura with their project.

5. CONCLUSIONS

OSS has several implications in education. For Education Management it can provide a means of obtaining better software at a cheaper price while encouraging the development of local industries and showing a better ethics model to students. In teaching it provides a model which can be shown to students on how to collaborate and develop projects between people who never get to meet on a face to face basis, but just using Information Technology tools. Considering the advantaged associated to OSS, we believe governments involved should try to promote OSS. Extremadura, a region in Spain, shows that governments can lead the road to implementing OSS.

6. FUTURE RESEARCH

Taking all we have examined into consideration, we believe a questionnaire could be developed to be sent out to those in charge of ICT in the Spanish Autonomous Communities and in other countries as well. The questionnaire should take into account if the Community has or has not developed OSS-related projects in the classrooms and should respond to matters such as the reasons that have lead to developing OSS, financial issues, development models, etc.

7. REFERENCES

Bulchand, J. (2005). Software libre en la Universidad. II Libro Blanco del Software Libre. Segovia, M.A. and Abella, A. Electronic book available at http://www.libroblanco.com. [Accessed September 30th, 2005]

Díaz, J.A.. (2005). Information Society in Extremadura. Information Technology and Educational Management in the Knowledge Society. Tatnall, A., Osorio, J. and Visscher, A. New York, Springer / IFIP: 171:180.

Golden, B. (2005). Succeeding with open source. Boston, Massachusetts, Addison-Wesley, Pearson Education.

Fink, M. (2003). The Business and Economics of Linux and Open Source. Upper Saddle River, New Jersey, Prentice Hall.

O'Reilly, T. (2006). Four Big Ideas About Open Source. Available at http://radar.oreilly.com /archives/2006/07/four_big_ideas_about_open_sour.html [Accessed August 23rd, 2006]

Pavlicek, R. (2000). Embracing Insanity: Open Source Software Development. Indianapolis, Sams Publishing.

Raymond, E. (2001). The Cathedral & the Bazaar. Sebastopol, California, O'Reilly & Associates.

Stallman, R. (1998). The GNU Project. Available from: http://www.gnu.org/gnu /thegnuproject.html . [Accessed September 30th, 2005]

Stallman, R. (2000) The GNU operating system and the free software movement. Open Sources: Voices from the Open Source Revolution DiBona, C., Ockman, S. and Stone, M. Sebastopol, California, O'Reilly & Associates. Available from: http://www.oreilly.com /catalog/opensources/book/stallman.html [Accessed September 30th, 2005]

Watson, R.T, Wynn, D. and Boudreau, M.C. (2005) JBoss: The Evolution of Professional Open Source Software. MIS Quarterly Executive, 4(3), Sept 2005: 329-341.

Weber, S. (2004). The success of Open Source. Cambridge, Massachusetts, Hardvard University Press.

Using LAMS to Link Learners in an E-Learning Environment

Learning Activity Management System

Maree A. Skillen
Shore (S.C.E.G.S), North Sydney, Australia

Abstract: As new technologies evolve and innovations emerge, educators' understandings about teaching and learning in the electronic environment are constantly challenged. Linking learners in both independent and collaborative settings becomes a significant pedagogical undertaking. A key component to gaining an enhanced understanding of this process involves the concept of Learning Design which has been described by Dalziel (2003b) as having "the potential to revolutionise e-learning by capturing the 'process' of education, rather than simply content". An example of a Learning Design system is the Learning Activity Management System or 'LAMS'. This paper will provide a brief outline of LAMS and a discussion that focuses on its use in a pre-service ICT teacher education course unit and with secondary students in a school-based Computing course. A sample learning sequence to be utilised by learners and developed for a selected content area will be briefly outlined. The aim for educators using LAMS is to create learning experiences that actively involve learners and improve educational outcomes. Early trials of this system suggest it allows learning and motivation to be enhanced and that there is a willingness exhibited by individuals to engage in tasks and whole group discussions. It is argued that teaching and learning can be enriched as the technological tools and pedagogical processes are brought together in appropriate ways.

Keywords: E-learning, LAMS (Learning Activity Management System), learning design, pre-service teachers.

1. INTRODUCTION

"Education is one of the fastest growing economic and social sectors in the world, and the use of new technologies is an integral and driving component of that growth".

(McCreal and Elliott, 2004, p. 115)

Please use the following format when citing this chapter:

Skillen, M.A., 2007, in IFIP International Federation for Information Processing, Volume 230, Knowledge Management for Educational Innovation, eds. Tatnall, A., Okamoto, T., Visscher, A., (Boston: Springer), pp. 123–132.

2. E-LEARNING

Whilst there is no universally accepted definition for *e-learning*, at its broadest, it is seen to be the use of ICTs (Information and Communication Technologies) to support teaching and learning. Siemens (2002) supports this by referring to e-learning as being 'the marriage of technology and education'. E-learning is often restricted to the online medium and is used widely to support different delivery modes in educational institutions.

Charles Clarke (DfES-UK, 2003) describes e-learning as having the potential to revolutionise the way we teach and how we learn. It is already present in schools, colleges, Universities, the workplace and in our homes. If someone is learning in a way that uses ICTs, then they are using e-learning. The concept of e-learning is important because individuals are finding and realising that this form of learning can make a significant difference in terms of how quickly a skill is mastered, the ease with which study can be undertaken and how much they enjoy learning (DfES-UK, 2003, p. 3). As Gerry White points out the benefits e-learning brings such as empowering the learner, providing choices and opportunities for collaboration across boundaries – make it an important player in 21st century education.

2.1 LEARNING DESIGN

Learning Design has emerged as one of the most significant recent developments in e-learning. Sloep (2002) refers to the teaching of Learning Design as being based on the general idea of people doing activities with resources. This shifts the emphasis from content driven single learner programs to more collaborative learning activity sequences. As Dalziel (2004) explains Learning Design is a name being given to a new field of e-learning technology based on 'best practice process'. He refers to it as providing "a glimpse at the ways of describing multi-learner sequences and the tools required to support these".

Learning Design can be considered to be a sequence of collaborative learning activities whereby each of the tasks can be stored, re-used and customised for alternative content or focus areas in a very short frame of time. In terms of the school context, Learning Design can be thought of as a series of lesson plans that can incorporate single learner content, but also collaborative tasks such as polling, forum, chat and small group discussions.

2.2 LAMS

LAMS (Learning Activity Management System) is a collaborative online learning system which was developed in 2002, by WebMCQ Pty Ltd in conjunction with Macquarie University. It allows for the establishment of a repository of learning materials where individual resources, tasks and constructed sequences to engage learners either independently or collaboratively, are stored for easy access or use by teachers and learners prior to activating them within a classroom environment through the use of ICT. Teachers have access to a facility for creating their own lesson activities or for reviewing deposited materials and then adapting these templates into alternative lesson sequences. This can be done quite quickly and numerous lessons can be generated from similar or very differently constructed learning sequences.

LAMS is a server-based application that can be accessed from any computer via a web browser. LAMS enables:

- Students to work independently and/or collaborate on-line through activities created using a tool set designed specifically for the purposes of quality on-line learning.
- Teachers to design learning sequences in accordance with sound pedagogical practices in a learning environment that can be enriched with local and web-based resources.

The LAMS application is divided into four levels of functionality (refer to diagram 1). These include:

1. The *Author area* is where teachers can design and modify learning activity sequences using specifically designed learning tools, such as discussion boards, research activities, group collaboration tasks.
2. The *Monitor area* provides a mechanism for teachers to assign learners to groups, activate learning sequences for specified groups, monitor individual progress through the learning sequence.
3. The *Learner area* accommodates individual and group learning sequences where students complete tasks in sequential order.
4. The *Administration area* is devoted to system administration tasks such as creating users and managing user groups.

LAMS 2 Architecture Overview

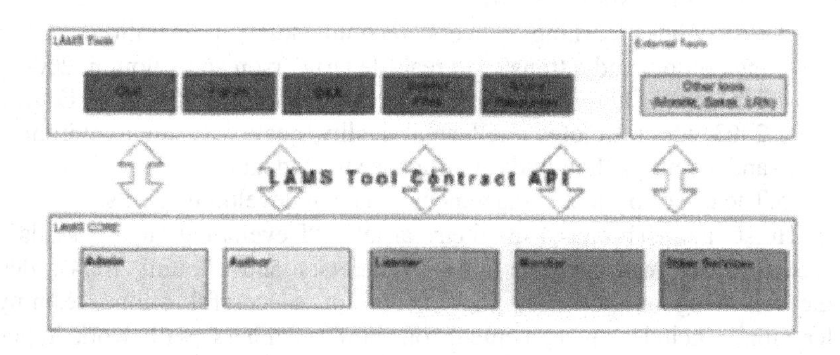

Diagram 1 (Dalziel, 2006)

2.3 USING LAMS WITH PRE-SERVICE TEACHERS

Pre-service Secondary ICT teachers' were required as part of an undergraduate teaching unit entitled 'ICT in the Secondary School' to develop a lesson using LAMS. In particular, students were required to demonstrate a basic understanding of LAMS and its relevance to e-learning; identify an appropriate syllabus and series of outcomes to use with a LAMS sequence; develop and implement a simple LAMS sequence; describe the process used to plan and implement a LAMS sequence; identify for the lesson/unit of work links to the 7-10 syllabus and, reflect on the importance of the learning process in the construction of a LAMS sequence.

2.3.1 A LAMS Lesson Example

Pre-service ICT teachers were required to become familiar with the NSW Board of Studies Stage 6 ICT syllabuses in order to identify a component related to legal, social or ethical issues that would benefit from the use of a LAMS interface as the lesson delivery mode. It is clearly identified that LAMS is well suited to "actively engage students in reflective activities that require them to work through ethical issues" (Meyenne, 2000), and LAMS can sustain a collaborative environment (when certain tools are invoked), so a group project within the Information Processes and Technology (IPT) and/or Software Design and Development (SDD) syllabus involving ethical decisions was sought. The school magazine/newsletter project from the Stage 6 Preliminary IPT syllabus was deemed as being the most suitable option for this task.

Having selected a component of the syllabus to be addressed, the next stage was to derive a lesson plan that best served the students' learning needs. The following were taken into account when planning the lesson:

1. Cognitive implications i.e. when students are required to use higher order (abstract) thinking, they require cognitive strategies and cognitive processing opportunities (Rosenshine, 1995).
2. Suggestions from Meyenne's methodology for the teaching of ethics in schools (Meyenne, 2000).
3. An online learning process recommending preparation, activities, interaction, and a transfer to real life (Anderson and Elloumi, 2004).
4. The NSW Quality Teaching Program (NSW Department of Education, 2003) that promotes intellectual quality, quality learning environments, and a context that holds significance to students.
5. The use of project management and team developing skills.

Hurst and Thomas, in their article "Developing team skills and accomplishing team projects online" (Anderson and Elloumi, 2004), identify the following as "the key ingredients for successful online teaming in learning" include an agreement on how members will work together, individual and group accountability, flexibility, the monitoring of progress, and social interaction. Each of these characteristics were incorporated into the assignment task developed for the specified course unit.

2.3.2 LAMS Activity Tools Selection Process

Rosenshine (1995) produced an article entitled "Advances in research on Instruction" in which he identifies six steps that best help students learn, particularly subjects that require higher-order thinking strategies. This article provided a framework for the development of a LAMS lesson, and the LAMS tools used in the Code of Ethics sequence were chosen specifically because they provided the functionality required for each step, as outlined below.

Rosenshine's Effective Teaching Strategies	LAMS Tool with the required functionality
Present new material in small steps.	Multiple noticeboards and Q&A are used to dissect information into manageable quantities.
Help students develop an organisation for the new material.	An example Code of Ethics is on hand for the teacher to show as a guide.
Guided student practice – provide opportunity for student processing.	The shared resources and polling activities give students cognitive processing opportunities.
When teaching higher order tasks, provide students with cognitive strategies.	The noticeboard tool provides an ideal medium for suggesting strategies when they are most needed.
Help students to use cognitive strategies by providing them with procedural prompts.	The polling activity is an ideal way to offer students a procedural prompt.
Provide students with opportunity for extensive student practice.	The progress mechanism allows students to revisit activities if needed. In this case, the extent of student practice is limited by time rather than process.

Table 1: Comparison of Rosenshine's Teaching Strategies and LAMS Functionality

3. FROM THEORY INTO PRACTICE

Many of the pedagogical practices espoused by the main schools of learning are incorporated in online learning (Anderson and Elloumi, 2004). In essence, the main learning theories can be summarised as: behaviourists favour teacher-directed learning sequences; cognitivists endorse the use of internalised learning and personal reflection and, constructivists encourage active rather than passive learning.

By way of distilling vast quantities of pedagogical theory and research, the NSW Department of Education have prepared a Quality Teaching Program (QTP) for NSW Public Schools. This model identifies three key issues to ensure quality student learning. The issues are intellectual quality, quality learning environment and, significance.

The "Code of Ethics" LAMS sequence developed by a student for the pre-service ICT teacher education assignment embraced each of the main learning theories as well as elements of the key issues identified by the Quality Teaching Program.

The teacher can ensure the intellectual quality of the lesson through:

- Articles and web pages to be reviewed or researched by the students are pre-selected and screened as being appropriate for the lesson.
- The sequence of learning activities is presented to students in a prescribed order. The activities are designed to meet specific learning objectives and the sequence ensures that pre-requisite knowledge is gained before subsequent learning activities are attempted (in accordance with behaviourist practices).

LAMS provided a tool to enhance the quality of the learning environment. This was demonstrated by:

- Individual contributions to group work could be monitored and assessed.
- Students were given the opportunity to self-regulate their learning by:
  - Conducting independent research of web-sources *in a supervised environment.*
  - Having time to reflect on materials presented (exemplifying cognitivist ideology).
  - Assuming responsibility to accomplish tasks within a group project *without teacher intervention.*

In accordance with constructionist theory, the project had inherent significance and encouraged active participation.

- The learning activities were directly linked to a scenario that was relevant to the school-aged students (that being the creation of a Code of Ethics to apply to the production of a school newsletter).
- Peer-to-peer and teacher feedback that recognised and gave value to all student contributions was encouraged.

In conclusion, The "Code Of Ethics" LAMS sequence provided students with the opportunity to explore, consider, discuss and synthesise ethics in an environment where each student could contribute, reflect and achieve both individually and as a valued team member.

4. DEVELOPING A LEARNING ACTIVITY SEQUENCE

To illustrate more specifically the implementation of Learning Design an example of a learning activity sequence, "Take a Stand", has been constructed to involve secondary IPT students collaboratively in considering issues, responding with an opinion and supporting their individual viewpoints. Table 2 summarises some of the key aspects to be considered in the planning process of a LAMS learning sequence.

LAMS Activity Title	Take a Stand
Year Level	Year 11*
Suggested Duration (Note: this will be influenced by the length of a lesson and the students application to the task)	4 to 6 lessons
Subject Area Focus	Information Processes and Technology: Preliminary → Social and ethical issues
Introduction to Task	As individuals we often agree and disagree on various issues depending upon our personal views to a given situation or topic. This activity will require you to consider a statement and to then 'Take a Stand'. There is no right or wrong, but individuals must be able to justify appropriately why they agree, disagree or are neutral about the selected statement. All opinions need to be supported with valid reasons by an individual learner.
Suggested lesson structure for completion of the sequence	Lesson 1: Initial Noticeboard + Polling Lesson 2: Chat & Scribe Lesson 3: Q&A + Read Noticeboard (#2) Lesson 4: Read Noticeboard (#2) + discuss Lesson 5: Submission + general discussion Lesson 6: Follow-up discussion

Table 2: Planning for a LAMS Sequence (Note: * the task shown could be adapted for younger students or made more complex to challenge older students.)

Table 3 summarises each of the LAMS activity tools used in 'Take a Stand' and aims to identify the application of each within the short sequence constructed.

Tool	Description of LAMS Activity Tool	Application of LAMS Activity Tool
Noticeboard	• Simple text to learner relating to the sequence and operation.	Introductory text for learners to put a context to the activity that they are about to engage in.
Polling	• Teacher provides learners with a list of options to "vote" on. • Shows collated learner responses.	Six issues will be listed separately for learners to read and to respond to: ie. they will vote or 'Take a Stand' (agree, neutral, disagree) for each of the issues in turn. Example: "We would have more unemployment if computers did not exist"
Chat & Scribe	• A live chat is combined with a scribe tool for collating the	Learners are posed with a series of questions to respond through

	chat group's view on a question created by the teacher. • Use in *small* groups: creates a parallel chat and scribe area. • Shows the outcome of each group on a whole class page (viewable by all learners).	collaborating in a group situation. The question may include, for example: "Many people believe that there is a danger that the computer technology revolution could be used as a tool by the wealthy industrial nations to further exploit developing nations. Do you believe there are ways in which wealthy nations could be preventing this to occur?"
Q & A	• Teacher poses a question to learners individually • Learner enters a response • Learners view all responses of peers on a single answer screen	Example question for learner to respond to may include "Computer users are sometimes able to gain unauthorised access to computer systems. What implications are there if this should occur?"
Noticeboard (#2)	• Simple text to learner relating to the sequence and operation	Your final task, to be completed for homework and then submitted in the next lesson, is to write an essay in response to the following statement: "Learning about computers should be compulsory in schools. Do you agree?" Your answer should state the reasons for your support or lack of support about the given statement. Explanations provided should be well discussed".
Submission	Learners submit a file for assessment by the teacher – for example, an essay or report Note: the monitoring workspace provides assistance to the teacher with managing the marking process for submissions.	Next lesson: learners should be able to submit a completed essay to the question stated in Noticeboard (#2). This homework item will be used for assessment purposes for the learner.

Table 3: Description and details of LAMS activity tools used in the sample sequence 'Take a Stand'

5. EARLY TRIALS OF LAMS

Beta versions of LAMS were tested widely in a variety of primary and secondary schools, colleges and Universities across Australia, Canada and the UK (Kraan, 2003). From these trials, preliminary findings have indicated that the system has assisted in increasing the participation rate of some students dramatically, along with enhancing the collaborative interactions of learners. Dalziel (2003a) found that initial evaluations of LAMS in K-12 schools and University environments during 2003 indicated that there was a profound impact on both student learning and teachers' conceptualisation of the learning process. Research into this impact is currently ongoing, but one early example from Kemnal Technology College, a boys' secondary school in the UK, found in a pilot evaluation that only 15% of students were willing to discuss ideas in front of their peers in the classroom, but over 80% of the same students were willing to discuss their ideas within LAMS (Gibbs, 2004). Interviews with teaching staff at Kemnal Technology College suggest that LAMS is one of the main contributing factors for their students becoming more active participants in classroom discussions and collaborative activities (Gibbs, 2004).

6. CONCLUSION

As with most things involving technology, there were limitations within the LAMS system that the design team have overcome in later versions of the software. The subsequent development of the tools and additional features within the system, in recent times, have added a greater depth and dimension to the LAMS experience for both learners and teachers. Whilst there may be drawbacks in educational institutions relating to the access and equity of ICTs, the advantages of using LAMS far outweigh the disadvantages. The advantages relate to the learning experiences of individuals where collaboration and the promotion of learner involvement are encouraged and cognitive skills are enhanced. The success of LAMS is such that it is now open-software and far more widely used in the area of education both within Australia and beyond.

The process of developing a unit of work in LAMS encourages teachers to reflect on their pedagogy and to find ways to engage their learners. They also have the option of using templates designed by other educators either as they exist or through easy adaptation to their own needs and specific subject areas. The use of LAMS in the classroom is a means of realising the potential held by technology for education in the 21st century.

7. REFERENCES

Anderson, T. and Elloumi, F. (2004). Theory and practice of online learning. Athabasca University: Canada. Retrieved March 21, 2004, from http://cde.athabascau.ca/online_book/index.html

Dalziel, J. (2003a, December). Implementing Learning Design: The Learning Activity Management System (LAMS). Paper presented at the ASCILITE 2003 conference, Adelaide, South Australia.

Dalziel, J. (2003b). LAMS: Teacher Guide – Draft. Macquarie University: MELCOE, Sydney.

Dalziel, J. (2004, January). The Learning Design Revolution: Implementing the Learning Activity Management System (LAMS). Paper presented at the Oxford Learning Design workshop, Oxford, UK.

Dalziel, J. (2006). Modelling a team-based astronomy task using LAMS. Presentation for ICALT 2006, Kerkrade, Netherlands, 5th July, 2006.

Department for Education and Skills-UK [DfES-UK]. (2003). Towards a Unified e-Learning Strategy: Consultation Document Executive Summary. Retrieved June 4, 2004, from http://ferl.becta.org.uk

Gibbs, D. (2004). Electronic selves: gender and anonymity as factors in e-learning. Paper to be presented at the Style Council Conference 2004. (Publication forthcoming)

Kemnal Technology College, UK. Retrieved June 9, 2004, from http://www.ktc.bromley.sch.uk/

Kraan, W. (2003). Learning design inspiration. Retrieved March 21, 2004, from http://www.cetis.ac.uk/content2/20031105152011/printArticle

McGreal, R. and Elliott, M. (2004). Technologies of Online Learning (E-Learning). In Anderson, T. and Elloumi, F. (2004). Theory and practice of online learning. Athabasca University: Canada. Retrieved March 21, 2004, from http://cde.athabascau.ca/online_book/index.html (pp. 115-135)

Meyenne, A. (2000, November). A proposed methodology for the teaching of information technology ethics in schools, Australian Educational Computing 15, 2 (pp. 15-20).

NSW Board of Studies (BOS). Syllabus Documents for Stage 6 IPT and SDD. Retrieved October 14, 2005, from http://www.boardofstudies.nsw.edu.au/syllabus_hsc/index.html

NSW Department of Education (2003). Quality teaching in NSW public schools: A classroom practice guide. Retrieved October 14, 2005, from http://detww.det.nsw.edu.au/directorates/profcurr/welcome.html

Rosenshine, B. (1995). Advances in research on instruction. The Journal of Educational Research, 88, (pp. 262-268).

Siemens, G. (2002). Instructional Design in E-learning. Retrieved March 21, 2004, from http://www.elearnspace.org/Articles/InstructionalDesign.htm

Sloep, P. (2002). IMS Learning Design Update. Retrieved June 9, 2004, from http://www.cetis.ac.uk/groups/20010809_144711/FR200211601_20327

White, G. (Ed). The changing landscape: e-learning in schools. Retrieved June 9, 2004, from http://www.educationau.edu.au/papers/changing_landscape_gw.pdf

Evaluation of a Web-Based Training System

for reading scientific documents based on activating visual information

Yukari Kato and Toshio Okamoto
Tokyo University of Agriculture and Technology, Nakach, Koganei, Tokyo, Japan
The University of Electro-Communications, Chofugaoka, Chofu, Tokyo Japan

Abstract: This study investigates the conditions under which graphical information can support reading comprehension in a second language. This paper presents results of an experimental study in which reading comprehension of foreign students in two different environments was compared: vocabulary learning only vs. vocabulary and grammar plus graphic tasks. The results indicate that tasks decreasing grammatical complexities were more effective than lexical practices.

Keywords: Web-based training, Japanese for Specific purposes, academic reading, graphical information.

1. INTRODUCTION

In recent years in Japan, there has been a rapid increase in numbers of foreign students studying at science and technology graduate schools. These students have already completed some academic courses in their major fields, but do not have language ability sufficient for academic life in Japan. In order to integrate such students into their language environment, language institutes and international student centers provide intensive language courses and arrange tutoring programs. For foreign students at science and technology universities, however, there is little time to enrol in regular Japanese language courses or to utilize available language learning opportunities.

To solve these problems, many universities and institutions have developed web-based language learning support systems that are free and open to all users. Most web-based systems are limited to automatically displaying the meaning of unknown words and the structure of each sentence by using morphological and syntactic information. Non-native readers use these systems mainly to decrease the time needed to look up

---

Please use the following format when citing this chapter:

Kato, Y. and Okamoto, T., 2007, in IFIP International Federation for Information Processing, Volume 230, Knowledge Management for Educational Innovation, eds. Tatnall, A., Okamoto, T., Visscher, A., (Boston: Springer), pp. 133–139.

unknown words in a dictionary. The validity of these systems has not been supported by language learning theories and psychological research.

In the fields of pragmatics and cognitive linguistics, there has been a renewed interest in "Relevance Theory" as proposed by Sperber and Wilson (1986). Sperber and Wilson (1986) assume that complex communication combines two different modes: *the coding-decoding mode* and *the inferential mode*. Based on this hypothesis, linguistic coding and decoding might involve the use of coded signals that fall short of fully encoding the communicator's intentions and merely provide incomplete evidence about them. In other words, communication is successful not just because readers are able to recognize the linguistic meaning of the author's arguments, but because readers also infer the author's true purpose in using them. Therefore, a *coding–decoding* process should be subservient to the *inferential* process (Silberstein, 1994).

Silberstein (1994) suggests that readers practice interpreting the function of various graphics, so that they can respond to them quickly and appropriately. However, it must be noted that the significance of the graphics is interpreted and explained only within the text (Moriarty, 1996). Therefore, readers need to understand not only individual details, but also the relationship among ideas for thorough comprehension of academic articles (Silberstein, 1994). Therefore, it is necessary to develop a relevant learning environment to manage the integration between graphical and textual information for academic reading comprehension.

So far, we have presented the hypothesis of communication proposed by Sperber and Wilson (1986). Based on their relevance theory, we shall later discuss our framework for a relevant learning system. Through implementing instructional devices based on the *Relevant Language Learning Framework* (Kato et al. 2002), this study provided empirical data used to identify factors that influence use of graphical information. The purpose of this framework is to propose a general guideline for Japanese language learning through use of experimental and theoretical data to identify information relevant to academic reading comprehension. According to the *Relevant Language Learning Framework* (Kato et al. 2002), four different reading strategy modules (lexical, grammatical, rhetorical, and graphical) were investigated during academic reading comprehension using computer-based instruction.

This paper is organized as follows: Section 2 describes the experiment conducted to investigate the influence of courseware design on use of graphical and textual information in academic articles. Section 3 outlines a proposed instructional model, based on both experimental and theoretical principles for an effective reading support system for non-native speakers, which facilitates the use and integration of graphical and textual information.

2. EXPERIMENT

In this experiment, the efficacy of two different courseware designs was tested by elaborating four reading strategy modules in the courseware. The most important difference for facilitating graphical information in reading comprehension occurred between the two grammatical and lexical modules.

2.1 Research Questions

The question at hand is whether courseware based on the *Relevant Language Learning Framework* (Kato et al. 2002) can facilitate reading comprehension in foreign students. Two types of courseware design were used to examine interaction between graphical modules and three strategic modules (lexical, grammatical, contextual). Two types of measurements (assimilative and discriminative questions) were used for reading comprehension. The following questions were investigated:

1. *How does the grammatical and lexical module influence graphical information use in the assimilative process of reading comprehension?*
2. *How does the grammatical and lexical module influence graphical information use in the discriminative process of reading comprehension?*

2.1.1 Subjects

The experimental subjects were 12 foreign students studying at national and private universities in Japan: 6 graduate students and 6 undergraduate students, including both intermediate and advanced learners of Japanese. Most participants had already received 1-2 years of formal instruction but were not yet able to pass the first level of the Japanese Proficiency Test. To examine the two different conditions, participants were randomly assigned to one of two courses. Five students participated in courseware A and seven students participated in courseware B:

Courseware A: (n=5; 2 undergraduates and 3 graduates)
Courseware B: (n=7; 4 undergraduates and 3 graduates)

2.1.2 Materials Structure and Procedures

Materials

We selected an article that appeared in a Japanese journal of information processing (Matsukura, 1999). The article describes advantages and disadvantages of a new meeting support system in comparison to previous meeting styles. The article had been used in prior research concerning the effects of graphical information on reading comprehension (Kato et al. 2001).

Design and Procedures

Two types of courseware design were used to examine the interaction between four reading strategies. They shared the following similar constructions: 1) pre-question, 2) reading strategic modules, and 3) reading passage with two types of reading comprehension tests.

Participants were tested individually on web-based courseware on the Learning Management System (http://conery.ai.is.uec.ac.jp). Each participant read the reading passages with strategic tasks and completed the assimilative and discriminative tests online. The access logs and participant scores were recorded on the WebClass server.

In the first section (pre-question) of the courseware, participants answered a questionnaire aimed at determining lexical knowledge related to the reading passage.

In the second section (reading strategic module), there was an important difference between the two coursewares on the weight of reading strategic lexical practices.

In courseware A, participants only answered a 27-item vocabulary quiz lexical module. In contrast, in courseware B, participants completed four strategic tasks: (a) vocabulary quiz (10 items), (b) grammar practice (9 items), (c) information transfer activities between graphical and textual information (3 items), (d) paraphrasing to identify main idea of discourse (5 items). Figure 1 shows examples of (c) information transfer activities.

In the third section, two types of measurements (assimilative and discriminative questions) were used to assess reading comprehension. Assimilative questions were prepared to examine immediate aspects of understanding (Widdowson, 1978). Thus, they involve the realization of prepositional and illocutionary value by reference to what has proceeded. Discriminative questions were also prepared in order to facilitate conveyance of the main message. They deal with relative significance, enabling us to take notes and write summaries.

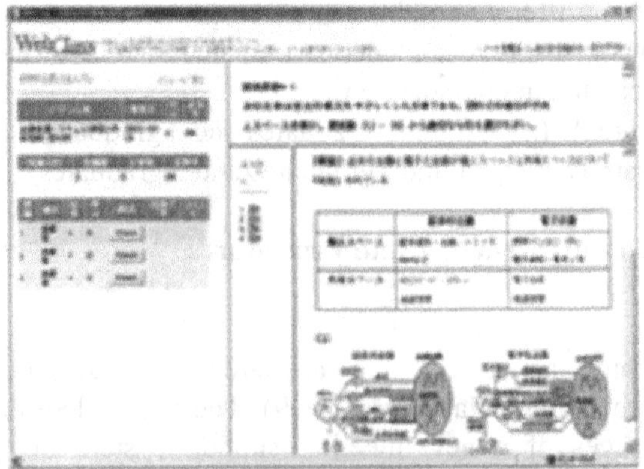

Figure 1: Information-Transfer Modules

3. RESULTS

We conducted two types of reading comprehension tests: assimilative and discriminative. The experiment aimed to examine the relationship between graphical modules and three strategic modules (lexical, grammatical, and contextual).

3.1 Correlation Analysis of Assimilative Questions

The lexical scores on pre-questions were positively related to the total scores on assimilative questions (r = 0.63). The correlation matrices reported in Table 1 were computed for each of the seven participants in courseware B. As predicted, the scores on both grammar and graphical modules were positively related to the scores on assimilative questions. Table 1 shows the strong relationship between grammar modules and graphical modules.

Table 1: Correlation Matrices for Strategic Modules and Assimilation

	Lexicon	Grammar	Graphic	Context	Assimilation
Lexicon	1.00				
Grammar	0.69	1.00			
Graphic	0.60	0.96*	1.00		
Context	0.86	0.32	0.20	1.00	
Assimilation	0.60	0.89*	0.96*	0.32	1.00

*p < .05

3.2 Analysis of Discriminative Questions

Participants in both coursewares A and B first summarized the main idea and then completed multiple-choice questions. The descriptive data from the summaries was analysed using a rubric developed to score the accuracy of relational meaning to written arguments (Muramono, 1992; Kato et. al. 2003):

1. Systematic/Central Conception; accurate structural knowledge with complete and accurate connection.
2. Partial/Peripheral Conception; structural knowledge with complete and accurate connection.
3. Duplicate Conception; at least one accurate structure shown, but connections between structures are inaccurate or missing.
4. Irrelevant; structural knowledge shown is inaccurate and irrelevant.

The rubric broke summary sentences into four categories of central, peripheral, duplicate, and irrelevant as shown in Table 2. In this coding system, central concepts were seen as more desirable than a mere listing of individual peripheral and duplicate concepts. The results, listed in Table 2, revealed that participants studying in courseware B indicated more central concepts than those in courseware A. This is especially noteworthy in the case of high-scored participants in assimilative tests. The opposite conclusion would hold for low-scored participants, which did not show the difference in discriminative tests.

Table 2: Summary Evaluation

Participant	Nationality	Assimilation	Characters sentences	Cen	Parti	Du	Irrel
A-5	Taiwan	23%	20 (1 sen)	0	0	0	1
A-1	Taiwan	55%	203 (4 sen)	0	0	4	0
A-2	China	77%	127 (3 sen)	0	3	0	0
A-4	Taiwan	88%	97 (2 sen)	1	1	0	0
A-3	Australia	100%	107 (4 sen)	0	1	0	2
B-2	Indonesia	30%	197 (3 sen)	0	1	0	1
B-3	Thailand	66%	15 (1 sen)	0	0	0	1
B-4	China	77%	12 (1 sen)	0	0	0	1
B-1	Indonesia	77%	96 (3 sen)	1	0	2	0
B-5	Korea	88%	97 (2 sen)	1	0	1	0
B-6	China	88%	90 (2 sen)	0	1	0	1
B-7	China	88%	83 (2 sen)	1	0	1	0

Note: 204 (4 sen) = 204 characters in 4 sentences, Cen: central, Part: Partial/Peripheral, Du: Duplicate, Irrel: irrelevant

4. DISCUSSION

Kato et al. (2001) suggested that illustrations embedded in articles appear to retard comprehension in the less skilled group although they enhanced comprehension in the group of superior ability. This drives us to the question of how graphical information can be made effective for academic reading. The central issue is identifying the knowledge necessary to understand the purpose and function of typical graphical information in academic articles.

Concerning the first research question, scores on both grammar and graphical modules were positively related to scores on assimilative questions. Analysis of courseware B indicates a strong relationship between grammar and graphical modules. This result suggests that grammatical knowledge can facilitate use of graphical information, in turn promoting assimilative understanding of written arguments. This implies that grammatical knowledge plays an important role in students' ability to use graphical information and in understanding descriptive ideas in arguments.

Concerning the second research question, the greatest difference in performance on use of graphical information between courseware A and B students manifested itself in grammatical information rather than lexical information. This appears to be because grammatical information is closely related to graphical facilitation for high-scored participants, which could promote understanding of main concepts.

From the viewpoint of language teaching, focusing on familiar graphical materials may furnish opportunities to capitalize on previous knowledge

(Silberstein, 1994). On the other hand, in order to use textual information adequately, it is possible that considerable knowledge of grammatical features is required.

5. REFERENCES

Kato Y. et al. (2001). "Analysis of the strategic activation of visual information in academic reading comprehension", Japan Educational Technology, Vol.25, Supple. pp. 155-160.

Kato Y. et al. (2002).A relevant learning framework for nonnative speakers: a proposal for integrating textual and graphical information in Japanese academic reading, Information and Systems in Education, Vol.1, pp. 70-79.

Kato Y. et al. (2003). The standards for texts comprehension based on discourse-structure analysis: comparison of summary between Japanese students and foreign students, The Journal of Japanese Language Education Methods, Vol, pp. 43-44.

Matsukura, R., Watanabe, S., Sasaki, K. and Okahara, T. (1999).A study of face-to-face collaboration support system composed of mobile PCs, Information Processing Society of Japan Journal, Vol. 40, pp.3075-3084.

Muramoto, T. (1992) Variation of Summary, Japanese Journal of Educational Psychology, Vol.40, pp.213-223.

Silberstein, S (1994). Techniques and Resources in Teaching Reading, Oxford, University Press, Oxford, U.K.

Sperber, D. and Wilson D. (1986). "Relevance", Blackwell, Oxford, U.K.

Widdowson, H.G. (1978). "Teaching Language as Communication", Oxford: Oxford University Press.

Absence Makes the Phone Ring Yonder

An end-to-end attendance recording and tracking system

Brian Pawson
Massey University, New Zealand

Abstract: In New Zealand schools there is rising concern regarding pupil enrolment and absenteeism. MUSAC, a business unit of Massey University, has worked with two other service providers to devise an end-to-end tracking system for pupil attendance. The process begins with teachers entering attendance information by mobile phone and ends when the database is automatically updated with information returned by a caregiver contacted by automated voice or text message. The system utilises web services with mobile technology and the latest development tools.

Keywords: School attendance, enrolment, early notification system.

1. INTRODUCTION

The New Zealand Ministry of Education currently has some concern regarding the number of students that 'disappear' from school enrolment. These disappearances typically occur when a child moves from one school to another whether it be at the same level (the family moves to a new location) or between levels (primary to intermediate, intermediate to secondary).

The reasons for these children failing to re-enrol in a new school are many and varied but, in some cases the caregiver either knows it has occurred and condones it or actually prevents the child from re-enrolling so they can help at home or assist with criminal activity.

To reduce this growing trend the Ministry has engaged accredited vendors in a pilot scheme which automatically notifies a Ministry server when a child changes to a leaver status. Basic details are deposited on the server and are only removed when that student is 'claimed' by the receiving school. The deposit and claim events are automatic and are managed using web services.

Please use the following format when citing this chapter:

Pawson, B., 2007, in IFIP International Federation for Information Processing, Volume 230, Knowledge Management for Educational Innovation, eds. Tatnall, A., Okamoto, T., Visscher, A., (Boston: Springer), pp. 141–144.

2. EARLY NOTIFICATION

As an extension of this work, vendors have worked with the Ministry to deploy an Early Notification System (ENS) that can automatically contact caregivers whenever a child is absent from school. Notification occurs via synthesised voice message to a landline phone or cell phone, by text message to a cell phone or by email.

Secondary schools confront greater absenteeism problems than primary schools due to the need for students to frequently move between classes. There are many opportunities for students to leave the school grounds. Late arrival can also be a problem. Therefore uptake of this new technology is greatest in secondary schools at present.

The system is reasonably simple. Most schools take an early roll call and quickly enter the resulting data into a computerised tracking system (by 10am all absentees have usually been entered). The attendance officer then initiates the ENS program which lists all absentees, including primary caregiver contact details. Generally all are selected for contact but if necessary selected students can be omitted.

The data is then sent via a web service to the ENS server which begins contacting caregivers using a pre-determined escalation sequence. Typically the system tries first to deliver a voice message to a landline phone. If that fails a text message is sent to a cell phone. Next a voice message to a cell phone will be attempted and finally, if all else fails an email will be sent.

The system will learn which method is most successful for each caregiver and will use that method first the next time. If contact is made the caregiver presses keys on their phone to indicate whether or not they knew about the absence.

After a pre-determined time, usually about 1 hour, the school computer requests from the ENS server all results from contact attempts. This data is then used to automatically update the school attendance database. The system will print mail merge letters addressed to caregivers who have been unable to be contacted.

The system is able to free up the school attendance officer to work on higher level strategies for improving attendance by removing the drudgery of repetitive phone contact.

The ENS server is operated by a private company that charges the school on a per-call basis.

3. GATHERING ATTENDANCE DATA

Presently there are a number of different approaches to gathering attendance data, though most schools handle it centrally by having one or two operators processing teacher roll returns which have been collected by the office runner. As technology and networks have advanced, more schools are requiring teachers to enter their own roll check data using a terminal in their class room.

4. ROLLTRACK

MUSAC has worked with another private company to improve the way in which the attendance data is gathered by enabling teachers to send their return by cell phone. This system is known as RollTrack

Early in the day, the attendance officer uploads to the RollTrack server the entire days set of class lists. Teachers then log on to the service and download their class roll to the cell phone. They can scroll through the list and, with a few key presses can identify the absentees. They submit this data back to the service which is polled every few minutes by our software. Changes are uploaded and the attendance database is updated automatically.

The technology provider for the cell phones has prepared an attractive plan which includes free peer-to-peer calling and reduced rates on personal calls among other incentives. The system enables teachers to submit roll returns from traditionally difficult locations such as the sports field, school trips and camps etc.

The system takes advantage of state of the art computer and mobile communications technology to provide an end-to-end attendance recording and notification solution.

5. GENERAL MESSAGING

The ENS system also provides a general messaging service. MUSAC is currently the only New Zealand vendor that has implemented software to enable access to this service. We have used the distribution lists in Microsoft Outlook to provide contact details. The user can enter a message in text form which can be delivered (with mail merge capability for the student name) by synthesised voice or text to a landline phone or cell phone.

This enables quick efficient messages to be delivered to nominated groups such as board of trustees, parent teacher committee, sports teams and coaches, parent support groups and so on. Or if the occasion demands, such as flooding or other emergency, all caregivers can be efficiently contacted while staff are free to go about other more useful business.

6. OTHER APPLICATIONS

As MUSAC progresses with development of these products we are seeking new services that can be made available to improve data gathering and supply of on-demand information.

One such possibility involves enabling PDA style mobile phones to send anecdotal information about students to a central database. This information would be available to the pastoral care practitioners in the school and could be delivered to a PDA phone or computer terminal on demand.

7. PILOT DEPLOYMENT

By December 2005 there were 10 schools installed with a pilot deployment. These schools provided feedback that enabled final honing of the product to suit a variety of situations and user preferences.

The school is able to install any one or more features of the package (RollTrack and/or Early Notification and/or General Messaging). The selection is generally based on financial considerations as each service provider charges an annual fee plus a volume usage charge.

MUSAC receives a small percentage of the volume usage charge, which is used to cover the cost of development and ongoing maintenance and support.

8. INITIAL FINDINGS

By May 2006 there were three schools making regular use of ENS and two sites using RollTrack. Another 3 schools are piloting the ENS system.

One of the RollTrack sites is uploading 40 class rolls per day in a full deployment and the other is still in trial mode.

There are six schools using the early notification system (ENS). These schools are very pleased with the results and the most active are sending 20 to 30 caregiver notifications per day.

Both systems represent a significant commitment by the school because there is a service charge which has the potential to ‘balloon’ if not carefully monitored. Despite this cost we have found that schools have a desire to use technology to improve attendance monitoring and information systems. The ability to more reliably inform caregivers about their child’s attendance is considered important.

A Development of Learning Management System for the Practice of E-Learning in Higher Education

Wataru Tsukahara, Fumihiko Anma, Ken Nakayama, Toshio Okamoto
Graduate School of Engineering (Educational Organization), TokyoUniversity of Agriculture and Technology, Japan Graduate School of Information Systems,The University of Electro-Communication, Japan

Abstract: From 2004 to 2007 our university has a three-year project called 'GP Project' which is part of a national project named "Selected Efforts of the Distinctive University Education Support Program" (Good Practice Project, abbreviated as GP Project), with the support of the ministry of Education, Culture, Sports, Science and Technology of the Japanese Government. In this project we are challenged to replace more than 30 existing courses with e-Learning. The courses must be seamlessly integrated to a course grade information system which has been already in use for years in the educational affairs section. For this integration we introduced a new Learning Management System which is easy to customize so that it can integrates course grade information and e-Learning information. We developed a Learning Management System (LMS) by modifying a commercial LMS WebClass with which is easy to add new functions. The first phase system has been started to work and now we are preparing to integrate the course grade information into the LMS. This paper describes our currently running three year university-wide project and then explains the functions of the newly developed Learning Management System.

Keywords: e-Learning, Learning Management System, GP Project, learner's feedback.

1. INTRODUCTION

From 2004 to 2007 our university has a three-year national support project named "Selected Efforts of the Distinctive University Education Support Program" (Good Practice Project, abbreviated as GP Project) (MEXT(2004), UEC(2004)). The theme of the project is "The Practice of e-Learning with mutual interaction for specialized courses." The project involves the whole Faculty and the UEC office. The Center for Developing e-Learning (CDEL) takes initiative in this project. In this project, we focus on:

Please use the following format when citing this chapter:

Tsukahara, W., Anma, F., Nakayama, K. and Okamoto, T., 2007, in IFIP International Federation for Information Processing, Volume 230, Knowledge Management for Educational Innovation, eds. Tatnall, A., Okamoto, T., Visscher, A., (Boston: Springer), pp. 145–152.

Use of Digital Portfolio: This is directly related to our theme. We plan to integrate course grade information currently stored in the Educational affairs section database with learner information (such as student ID etc.) stored in the LMS. It enables teachers to give students appropriate suggestions. We currently plan to include:

- subject of essays and submitted essays
- status of submission and submission dates
- status of credits
- grades
- past history of mentoring with summaries

We will start this function from the second term after the current system is integrated with the grade database.

Mentoring and Coaching: Preparing a digital portfolio is not sufficient. It is necessary to provide the function of mentoring/coaching. Because this function is new to teachers of the UEC, we decided to have tutorial for teachers. Mentoring/coaching will be about:

- questions about contents
- recommendations of courses for the learner
- know how of proceeding with learning
- providing curriculum suitable for the learner

These comments are made in reference to the learner's digital portfolio. By appropriately supporting learners in mentoring/coaching, we believe that even the specialized courses can be learned by e-Learning.

Currently, we have 13 contents ready for e-Learning. Topics include business administration, computer literacy, cryptography, image processing, bio-informatics, media literacy, micro-systems, fluid dynamics, software design, fuzzy systems, artificial intelligence, and communication technology. There are various styles of contents:

- video recording or studio recording of the actual lecture
- text and image content written in HTML
- simulation based contents (Active X, Java applet)

Some teachers plan to carry out full e-Learning, and others plan to use it in blended learning style.

In this project we are challenged to replace more than 30 existing courses with e-Learning. In order to realize the above features, a tight connection is important between the LMS in CDEL and the course grade information database in the educational affairs section. The courses must be seamlessly integrated to the course grade information system which has already been in use for years in the educational affairs section. For this integration purpose we needed to introduce a new Learning Management System which is easy to customize so that it can integrates course grade information and e-Learning information. We developed a Learning Management System (LMS) based on a commercial LMS *WebClass* for which is easy to add new functions. We named this LMS WebClass-RAPSODY.

In the following section we briefly describe the systems we have previously developed to realize our aim in e-Learning and then introduce the new LMS and its function.

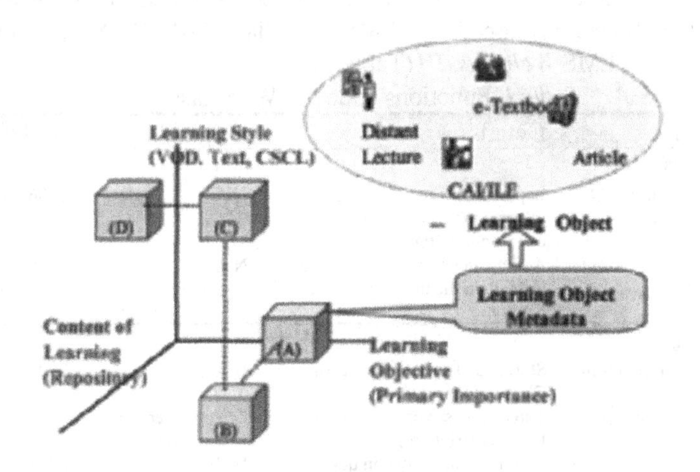

Figure1. Learning Ecological Model

2. BASIC IDEA: RAPSODY LMS SYSTEMS

An efficient e-Learning system provides a learning environment which has a high degree of freedom (ALIC (2003), Collins (1999)), letting learners choose appropriate learning contents. The Learning Ecological Model focuses on learning content, objective, and style (Seki et al. (2000, 2002), Okamoto et al. (2001)). RAPSODY is based on this model. On the other hand, teachers want to check access status for improving their course. To realize this, LMSs must handle information such as user identity, curriculum, learning contents, learning history, communication among learners, questions, and answers (Seki et al. (2002), Okamoto et al. (2001), Ymakita et al. (1999), IMS/GLC (2005)). RAPSODY systems (Seki et al. (2000, 2002), Okamoto et al. (2001)) implemented these features. In RAPSODY, learners and teachers can utilize functions such as computer supported cooperative learning (CSCL), authoring, planning curriculum, evaluation, and reporting. Contents are stored with Learning Object Metadata (LOM) proposed by IEEE-LTSC (2005) for the reuse of content. ADL SCORM (ADL (2004)) will be implemented in the next phase. Learner evaluation is done based on learning history.

There are external modules for sequencing (Seki et al. (2000)), CSCL (Kotani et al. (2004)), and plagiarism (Seki (2004)).

3. WEBCLASS-RAPSODY

Although RAPSODY has very advanced features, it is lacking in robustness. So we developed new LMS WebClass-RAPSODY by modifying a commercial LMS *WebClass*™ (Table 1).

Table I. Functions added to WebClass system

Function	Details	Original	Available
Administration			
Importing Student ID and course grade	Importing ID and course grade information from database in educational affairs section	N.A.	2006
Learner Management in UEC structure	Adding, deleting, and analyzing learners in the unit of year and faculty.	N.A.	2005
Authoring			
Content Management	Storing and searching of Learning Object for ease of reuse.	N.A.	2006
Structure of Contents	Three layer structure (Course , Unit , description)	Course, Scenario, Description	2005
Variety of question style	n-to-n mapping style in question	Multiple choice, single choice, text input	2005
SCORM 1.2	Enabling authors for annotating sequencing order.	N.A.	2006
Communication			
Mailing List Service	Sending mails among learners of the course.	N.A.	2005
Discussion board	Discussion board for each course	N.A.	2004
Mentoring Support	Digital Portfolio function both available for learner and mentor	N.A.	2005
Analysis			
Statistics	Histogram for standard deviation, average, highest, lowest scores in the assignments.	Average, highest, lowest scores	2005

Administration

When registering learners, teachers can choose them from an already registered partition which was imported from educational affairs section database. The partition has the usual tree structure so that the learner can be specified by regular/irregular course student, entrance year, faculty, and department (Figure 2). Importing course grade information is available from 2006 in XML format including student ID, name, entrance year, faculty, department, e-mail address, currently taking courses, past courses with grade, course's teacher ID, and teacher name. Using this information the student will receive various suitable suggestions from his learning history.

Authoring

A problem frequently pointed by teachers is the low variety of question formats at the end of each unit. We added n-to-n mapping question style.

Authors have to follow the structure of Course-Unit-Description/Question. Figure. 3 show a support for construction of a Unit from Descriptions (left top) and Questions (left bottom). Teachers choose Description/Question into Unit sequence (right) so that Descriptions appear alternately.

Communication

Supporting mentoring activity is one of our primary focuses. We call the support framework the Digital Portfolio. This can be an asset for a learner to reflect his progress in a course, and in the weekly schedule. Necessary information for the Digital Portfolio for a learner is: his one week course schedule, progress and ranking in a specified course, history of his progress (progress curve) including attendance, submission of essays, past mentoring comments, and submitted reports. Also, mailing list and discussion board can be used in mentoring/coaching.

Analysis

Teachers want to analyze their course during e-Learning period. In WebClass RAPSODY teachers can obtain basic statistics such as sum, average, maximum, minimum, standard deviation of access time and score. Using this information, teachers analyze their learners and courses (Figure 5).

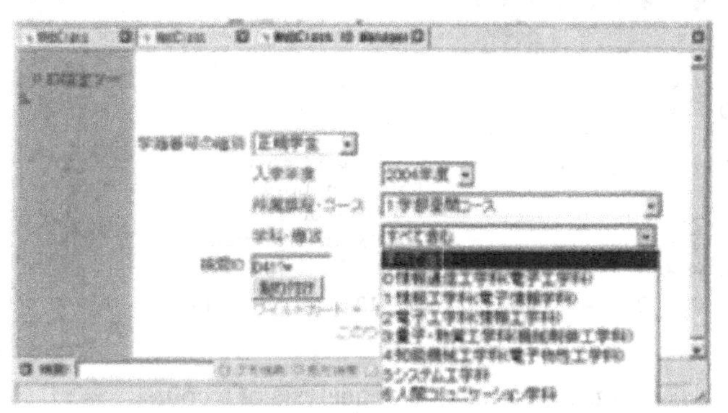

Figure 2. learner's specification

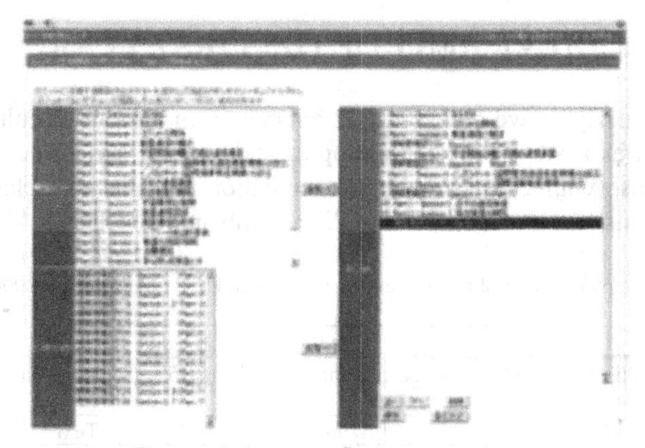

Figure 3. Support for construction

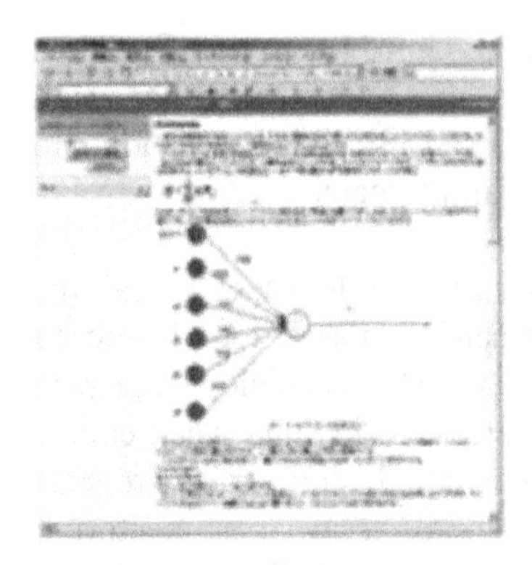

(a) WBT style (b) VOD style

Figure 4. Example of stored contents

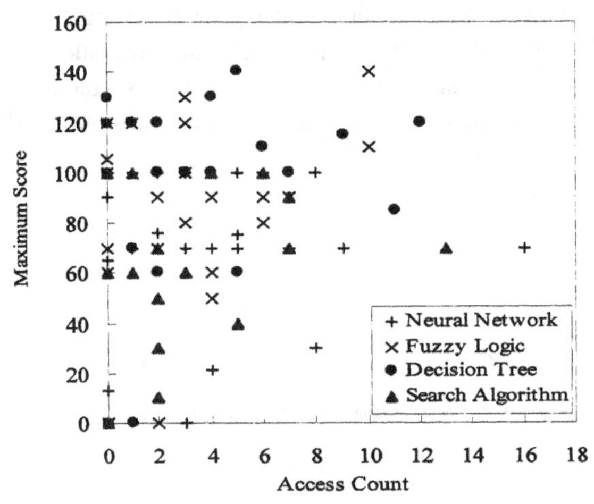

Figure 5. Example of analysis of a course

4. ACHIEVEMENT OF THE FIRST YEAR

As the first year we developed 13 courses from various Faculties (Table 2). There was a variety of course topics and content styles, not only with html, but also with video stream and simulation plug-ins. We plan to add 9 more contents and prepare a total of 22 contents in WebClass-RAPSODY.

Table II. List of the e-Learning contents in UEC GP Project in 2004.

Course Name	Semester	Units	Style
Artificial Intelligence and Knowledge Processing	Summer	15	Text/Video
Knowledge and Media design	Winter	8	Text
Instrument Technology	Winter	2	Text/Video

Space Communication Engineering (Lecture and Practice)	Summer	8	Text/Video/Simulation
Advanced Information Technology for Small Businesses with the Open Source Software	Summer	8	Text/Video
Advanced Image Engineering	Winter	10	Text/Video
Numerical Fluid Dynamics	Summer	12	Text/Simulation
Software Systems Design	Summer	10	Text/Simulation
Media and Systems II	Winter	8	Text/Simulation
Cryptography	Summer	12	Text/Video
Media Literacy	Winter	4	Text/Video/Simulation
Computer Literacy	Summer and Winter	13	Text
Exercise in Life Informatics	Winter	8	Text/Simulation

5. CONCLUSION

In this paper we introduced the UEC GP Project and the Center for Developing e-Learning (CDEL), which systematically stores e-Learning contents. For this purpose we developed WebClass-RAPSODY LMS. The system can support registration, authoring, and importing information from the educational affairs section. This enables effective mentoring/coaching by the use of Digital Portfolios.

Currently we have 13 contents from all over the Faculty. These contents contain various topics. In the next year we plan to evaluate how university education would change by this university-wide e-Learning practice.

6. REFERENCES

ALIC (2003): e-Learning white paper (in Japanese), Ohmsha, Japan.

Collins, B(1999): Design, Development and Implementation of a WWW-Based Course-Support System, Proc. 7th International Conference on Computer in Education, Chiba, Japan, pp. 11-18.

MEXT(2004): Distinctive University Educations etc, Support Program (Good Practices): http://www.mext.go.jp/english/news/2004/05/04052401.htm

IMS/GLC(2005): Global Learning Consortium: http://www.imsproject.org/ .

IEEE LTSC(2005): IEEE P1484.12.3/D8, 2005-02-22 Draft Standard for Learning Technology - Extensible Markup Language Schema Definition Language Binding for Learning Object Metadata : http://ltsc.ieee.org/wg12/files/IEEE_1484_12_03_d8_submitted.pdf

Kotani, T. et al(2004): Development of Discussion Support System based on Values of Favorable Words Influence, Journal of Japanese Society for Artificial Intelligence, Vol. 19, No. 2, pp 95-104.

Okamoto, T. et al(2001): The Distance Ecological Model to Support Self/Collaborative Learning in the Internet Environment, Journal of Educational Technology Research, Vol. 24, pp 21-32.

Seki, K. et al(2000): Construction of the Distance Teacher's Self-Training System based on the School Based Curriculum, Journal of Japanese Society for Information and Systems in Education (in Japanese) , Vol. 4, No. 3, pp 307-318.

Seki, K. et al(2002): an Adaptive Sequencing Method of Learning Object in e-Learning, Proceedings of JET02-1, Japanese Society for Information and Systems in Education (in Japanese), Iwate, Japan, pp 61-66.

Seki, M(2004): Development of Format Free Report Evaluation System in e-Learning Environment, MS Thesis, Graduate School of Information Systems, University of Electro-Communications, UEC, pp 1-64.

UEC(2004): UEC GP Project, Project Homepage: http://www-gp.ai.is.uec.ac.jp/

Ymakita, T. and Fujii, T(1999): A Hypertext Database for Advanced Sharing of Distributed Web Pages, Proceedings of 15th International Conference on Data Engineering, Chiba, Tokyo, pp 99-100.

WebClass(2005): http://www.webclass.jp

ADL(2004): ADL SCORM 2004 Conformance Requirements Version 1.2: http://www.adlnet.org

ITEM Everyday

Greg Baker
Scotch College, Melbourne, Australia

Abstract: This paper argues that the use of information technology in educational management (ITEM) is an important component of the efficient and effective operation of a large K-12 school. Systems in place allow the development of a complete picture of a student's activities and teachers are better able to make decisions that helps each student reach his potential.

Keywords: Educational management, school information system, student attendance, student reporting.

1. INTRODUCTION

Without the use of Information Technology in educational management, our school simply cannot function.

Every teacher uses the Scotch Information System every day. Similarly, most of the non-teaching staff also use the system.

Discussion at the 2006 ITEM Conference seemed to highlight the promise of information technology in this context, but there were significant reservations about the current advantage of such technologies in educational institutions.

This paper asserts that the use information technology is of vital importance in the ongoing effective and efficient operation of a large school.

2. BACKGROUND

Scotch College is an independent K-12 school for boys located about 7 km east of the centre of Melbourne. There is an enrolment of 1,850 boys including some 165 boarders. It is the oldest school in the state of Victoria and one of the oldest in Australia.

Please use the following format when citing this chapter:

Baker, G., 2007, in IFIP International Federation for Information Processing, Volume 230, Knowledge Management for Educational Innovation, eds. Tatnall, A., Okamoto, T., Visscher, A., (Boston: Springer), pp. 153–157.

It has a reputation for academic excellence but is also prominent in sport, music and the services. It has a broad curriculum and each boy is encouraged to achieve to his potential in a range of activities.

3. SCOTCH INFORMATION SYSTEM

The Scotch Information System had its roots in the late 1970s and was first developed in the early 1980s. A more complete account of the development of this system appears in the 2004 ITEM Conference proceedings (Baker 2005).

Since that time, there have been significant upgrades to provide a more complete picture of a student's activity within the school.

The goals of the Scotch Information System include providing:

- Accurate, up to date and relevant data
- A web interface that is easy and intuitive for staff to use
- Access from different locations including staff desks, classrooms, outside the classroom on the campus, at home and from a conference outside the school
- Different access rights to the data depending upon the need
- Integration with other information systems across the school
- A complete picture of a student's activity within the school and that stays as a permanent part of his record.

4. BREADTH OF THE SYTEM

The system is based on the person as the central entity. A person can have many attributes. Once a person is created on the system, the record is stored permanently with the exception of those involved in the enrolment process that ultimately do not attend the school.

The system includes information about:

- Students
- Staff
- Alumni
- Parents

Relationships between people are captured in a structured manner to enable tracking of families over time. The concept of the "Scotch Family" is an important one within the school and we seek to develop and enhance life long relationships with those who move through the school.

5. COMPREHENSIVE REPORTING ON STUDENT ACTIVITIES

A comprehensive student reporting system provides teachers with the opportunity to get a more complete view of a student's activities. These reports include:

- Academic progress report (four per year)
- Academic subject report (two per year)
- Tutor report (two per year)
- Sporting team report (at least two per year)
- Music group reports (two per year per group)
- Musical instrumental report
- Activity report eg debating
- Service report eg social service, cadets (two per year)
- Boarding report (two per year, boarders only)

These reports can be accessed by teachers for any of the students with whom they interact. Reports can also be generated in a range of different groupings as PDFs for either printing or reading on line.

The reports retain links to the person record. Hence, these reports are available on line indefinitely removing the need to archive them.

Sporting teams are often used for alumni reunions. Once a student report is created for a sporting team, that entire team or the individual report can be accessed as required. Included is the facility to add photographs of teams or individuals.

6. STUDENT ATTENDANCE

The accurate recording and reporting of student attendances is becoming an important aspect of school life. Brian Pawson has reported on the system his organisation has built in New Zealand "to provide an end-to-end tracking system for pupil attendance." (Pawson 2006).

Whilst the concern about truancy as reported by other schools is not significant, it is important to know when a student is at school and to ensure that absences are followed up and accounted for.

The system that we have in place allows teachers to record student attendances in the classroom. These attendance records are immediately available throughout the school for all to see.

Each teacher is allocated a notebook computer and has a personalized logon to the information system. Access can be wired or wireless – a virtual private network provides security for wireless access.

The process involves the following:

- The teacher logs onto the system. This logon is personalized and his or her class is displayed.
- The teacher records any absentees and saves the information.

- Secretarial staff can check which classes have not had attendance information recorded. In these cases, the teacher is sent a reminder message via the use of a pager.
- Secretarial staff enter any absence information that has been received by the school by email, telephone or voicemail.
- Tutors follow up unexplained absences to ensure that all absences are displayed with an appropriate reason.

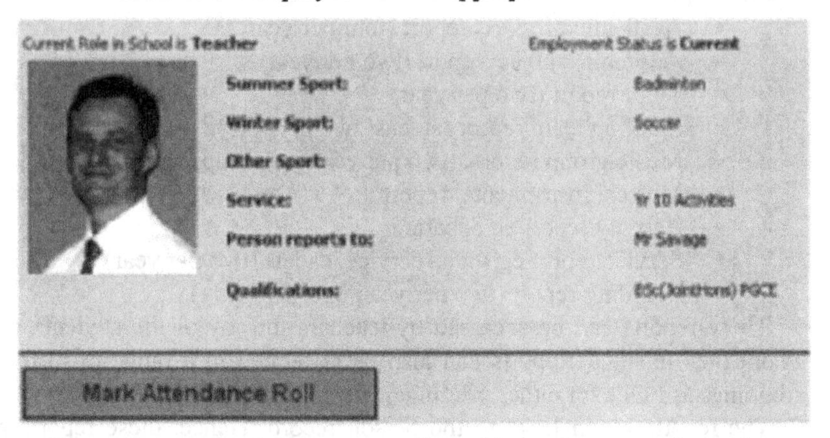

Figure 1 A teacher's logon page allows him or her to easily mark the attendance roll.

The system allows for a range of contingencies including:

- Students at excursions
- Students taking music tuition lessons
- Students involved in sporting activing
- Students away on camps.

The system provides:

- Accurate, up to date information
- Immediate feedback.

A range of reports on attendance and absence are available at the user's discretion.

7. TEST MATERIALS

A range of data gathered from external tests is available for use. For example, state wide AIM tests are conducted at Years 3, 5, 7 and 9. These provide scores on reading, writing and mathematic ability and can be compared to state and internal norms.

Other external test data is being collected and displayed. This type of data gives teachers the opportunity to compare "actual" versus "potential" on test scores and to better identify students who are under achieving.

8. KEY CHALLENGES

The volume of data is ever expanding. It is an ongoing challenge to be able to collect, store, manipulate and make sense of this data. Similarly, new standardized tests are being undertaken and government requirements on reporting are ever changing.

A key challenge is to be able to devise methods of determining when a student appears to be under achieving and to allow teachers to intervene to help that student achieve his potential. At the same time, the external data gives the school the opportunity to better understand the extent to which it adds value i.e. the extent to which students can achieve beyond their externally assessed potential.

9. CONCLUSIONS

The use of information technology in managing the school allows us to:

- Run the school effectively and efficiently
- Accurately track students' attendance in a timely manner
- Provide a more complete picture of each student's involvement in activities beyond the classroom.

In short, teachers are able to make better decisions about individual student progress and help students achieve their potential across a range of activities, both academic and co-curricular.

10. REFERENCES

Baker, Greg (2005) Developing an Integrated School Information System, Information Technology and Educational Management in the Knowledge Society, Tatnall, A. Osorio, J., Visscher A. (2005) Springer Science & Business Media, USA

Pawson, Brian (2006), Absence Makes the Phone Ring Yonder: An end-to-end attendance recording and tracking system. Published elsewhere in this volume.

Mapping the Future of Research in Web-Based Education Management

Effective HCI Taken to New Horizons

Elspeth McKay, Brian Garner, Toshio Okamoto
RMIT University, School of Business Information Technology, School of Engineering & IT, Deakin University, University of Electro-Communications, Tokyo, Japan

Abstract: The educational research community attracts practitioners and policy makers interested in both consuming and producing high quality educational research methodologies. Over time, research findings significantly contribute to continuing educational theory as well as educational management and professional practice. Until now researchers have been able to understand the interactivity of instructional strategies and cognitive performance in traditional learning models. While the community has been exposed to these findings through professional publications and the electronic media; with the advent of Web-based educational research, the same cannot be said about dissemination from this emerging techno-educational paradigm. Is this due to the complex nature of the contributing factors involved with educational management that involves information and communications technologies (ICT) in the Web-mediated learning environments? In terms of the relationship between information technology and education management, defining such educational research is becoming quite difficult. This paper analyses research by practitioners primarily interested in Web-based education. To examine the trends, contributions are examined from a wide range of educational researchers at a recent international conference that attracted participants from 29 countries. These trends will be extrapolated to map the future of Web-based education management based on international synergies in research communities of practice.

Keywords: Web-based, Web-mediated, information and communications technology, human-computer interaction, context-awareness, learning reinforcement, flow experience, podcasting.

1. BACKGROUND

Each year there are a number of international conferences that concentrate on educational research; none are actually convened to examine

Please use the following format when citing this chapter:

McKay, E., Garner, B. and Okamoto, T., 2007, in IFIP International Federation for Information Processing, Volume 230, Knowledge Management for Educational Innovation, eds. Tatnall, A., Okamoto, T., Visscher, A., (Boston: Springer), pp. 159–166.

education management. To this end, this paper reports on one such event, which took place in Melbourne, Australia, during November 2004; the International Conference on Computers in Education (ICCE2004-Full-Proceedings 2004). A unique feature of this conference was its concentration on high-quality research dedicated to educational technology in the Asia-Pacific region. Overall, this event attracted 350 registrations from 29 countries, with a total of 252 papers that were presented in the 3-day scholarly programme. This paper aims to provide a critical analysis of the scholarly discourse emanating from the papers that were submitted to this international conference. Leaving aside the notions of education management, the diversity of the educational technology community is shown first, followed by an analysis of the evidence from this forum on Web-based educational research, which suggests the future direction of research in Web-based education.

2. CROSS CULTURAL DEMOGRAPHICS

Overall there were 340 submissions representing a truly global knowledge exchange. Papers were submitted to a double blind peer review process. Where there were inevitable instances of disagreement and/or clash of review outcome, those papers in dispute were then submitted to a third independent review. Through this rigorous blind review, only the best quality papers were chosen as long papers. Examination of the Conference Proceedings reveals that the Reviewer Listing totals over 70 international reviewers. Finding suitable academic reviewers for a conference of this size would present a daunting task for any conference organizing committee. Figure 1 depicts the authorship demographics showing the global nature of the participation from the Asia-Pacific region and the wider community of educational technologists.

3. WEB-MEDIATED EDUCATION MANAGEMENT

3.1 Adaptive online tutoring and intelligent learning tools, and virtual reality

There were 42 papers that dealt with advanced HCI to deliver computer-generated intelligence. An examination of the authors' affiliation reveals unilateral research team membership where there may be little opportunity to collaborate across multidisciplinary fields. Perhaps this is not so surprising, according to Preece (1994) the major contributions to HCI have come from the more traditional fields of computer science, cognitive psychology, social and organizational psychology, and ergonomics and human factors. However it is interesting to note that it is the *virtual reality* research projects presented in these conference proceedings that combine multi-skilled teams, to produce an interesting synthesis of professional practice. Whereas the *adaptive online tutoring* projects, do not. Instead they have a strengthened focus on the technological development per se. While these latter mentioned research teams are succeeding with their projects, they must remain cognizant of the danger of not considering the human-dimension of their HCI. This means keeping one eye on the technology of their system building, while at the same time making sure that they are mindful of how this system fits in with what people are doing with this system (Peece, 1994).

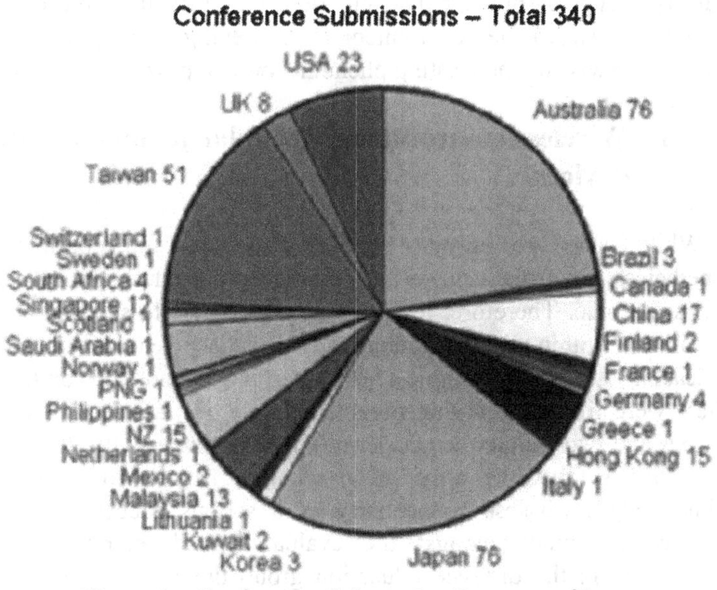

Figure 1 : Conference Submission Demographics

3.2 Collaborative knowledge sharing, teacher education, language learning, and assessment techniques

There were 78 papers in this group that focussed on knowledge acquisition. They represent the largest number of projects that implement ICT in a Web-based environment. Multi-disciplinary collaboration is strongest amongst the education and information technologists. However, notable amongst this contribution to education management and information technology research is the paper on *context-awareness* (Li, Zheng, Ogata and Yano 2004). While Li et al (2004) concentrate on the portability of learning environments and do not refer directly to the practice of Web-based instruction; there is a clear message on (education) *knowledge management* for researchers to learn from here. That is, the requirement for researchers to consider the social nature of learning. This important inter-relationship between *tacit and explicit knowledge* is captured by these researchers with their five-dimensional representation for context-awareness.

The problem of *knowledge retrieval* permeates through many of these papers. The advent of multimedia means our propensity to become knowledge squirrels is realized. Visual image storage has brought forward this dilemma. This is a problem that is growing rapidly due to the increase in the range of affordable digital technological devices and the ease in which the information/knowledge is captured as textual/visual objects. There are many untapped educational resources that can enhance a student's understanding. To this end, high quality database management repositories are critical. As far as education management is concerned, research from the industry sector is showing that attention must turn from the concentrated textual retrieval to the requirement for retrieving stored images from the www. Otherwise the podcasting phenomenon will pass education sector by.

3.3 Wireless environments (cellular phones and PDA devices)

While there are only 14 papers that focus on portable devices, nonetheless they offer a powerful reminder to this educational forum of the things to come. Therefore, it comes as no surprise to see this novel work positioned within the community of distance education researchers. Podcasting is a prime example of the emerging technologies on the horizon. There were five projects that involve *cellular phones*, while two use *PDA devices* and two devote their paper to wireless connectivity. It is interesting to note here that the only other scholarly stream of research which acknowledges wireless connectivity as a serious education management resource is found amongst the evaluation and learning technologies proponents. In the distance education group there were five papers from Japan, and two from Taiwan that focus on the technology-dimension of these mobile devices. However, it is in the contributions from Australia and the UK where the human-dimension is identified strongly.

4. SUMMARY OF VIEWS

Seven conference sub-themes covered a wide variety of scholarly ePedagogy innovating effective human-computer interaction (HCI) through ICT. Papers examined showed evidence of an explicit relationship with Web-based learning, while others showed a more implicit connection to a Web-mediated educational context. Four categories of research were identified: *differentiation capability for HCI*, *knowledge management*, *transportable environments*, and *design*. In the first category, the majority of the work fell amongst the AI/networked learning and the evaluation community. The papers that were strongly aligned to knowledge management attracted a broader coverage; these papers surfaced through the distance learning, eLearning, and Web-mediated learning tool conference stream/topics. While papers on transportable educational devices were aligned to distance education, they were the smallest concentration of Web-based research. Finally, there were 38 authors who displayed a connection to the process of design for Web-based learning; these papers were represented in the widest range of conference topics.

Differentiation capability for HCI: The highest proportion of papers for this group are to be found in the online tutoring, multi-agent social learning simulations, and assessment techniques. The work in this category concentrates on the human-machine dimension of HCI. Many of these papers describe a type of computer-generated intelligence that is generated by their learning environments. Understandably, these adaptive systems require the expertise from computer science or electronic engineering to predict the behaviour of the ICTs. Techno-knowledge about software design and development is critical with these projects. Meanwhile, excitement surrounding the virtual reality world brings about new synergies between the cognitive scientists who are primarily interested in the human-dimension of HCI and the machinery experts. This emerging field interested in various types of Web-based educational virtual worlds can be found in four of the seven conference themes. These papers all have the common characteristics of virtual reality because they allow the participant's senses to seemingly interact with the system. In virtual reality these systems should provide the participant with a sense of direct physical presence, some type of sensory cue in 3D (sight, sound or touch), and a natural feeling of interaction (Preece, 1994). However, it is the papers emanating from the conference themes of Evaluating Teaching and Learning Technologies that the human-dimension of HCI leads the way forward. As such, these papers concentrate on the cognitive aspects of Web-based teacher educational platforms to deliver learning/instructional systems for reading and language learning.

Knowledge management: By far the heaviest concentration on Web-based knowledge management is found in the conference themes of Distance Education, eLearning and Knowledge Management, and Web-Mediated Learning Tools; while, there are a small number of papers in the AI and Evaluating Teaching and Learning Technologies conference sub-topics. However, this is the body of research that serves as the vital link between the realms of human-machine fit and adaptation, human-dimension, and the development process for Web-based interaction. Moreover, this is where the three important elements of learning are implemented; they include: the *skills* that are needed for task-related activities that are necessary for humans to perform at basic levels, *knowledge acquisition* we need to provide a deeper understanding of underlying concepts, purpose and functionality with the whole system in both the operation and conceptual level, and *attitude development* which underpins effective skill management (Smith, 1997); (Gagne, 1985).

Transportable environments: Analyzing the scholarly activity in educational technology research has revealed the emerging interest in wireless technologies. At this point, within the ICCE community, the contributions appear to be only tinkering with such devices. This is where a fresh approach to Web-mediated learning is initiated with communications engineering expertise. These transportable environments appear to offer major advancement in educational technology for the future. However, more research is required from the educational technology community before the effectiveness of these ICT tools is known. It also important to note that we must learn from the past, and ensure there is multi-disciplinary collaboration. Otherwise the negative effects of the so-called digital divide will restrict these innovations.

The danger for these transportable environments within the education/training sectors lies in our lack of understanding how to overcome the tendency to work at an ever increasing pace. Flicker (2002) describes four rules to overcome this tendency; they apply to educational technologists as much as they do to the corporate sector. These rules are to; avoid confusion by adopting inclusive design strategies, make a decided shift from racing to pacing, work within people's natural zone or flow state, and fulfill both people and the expected project outcomes. The concept of *flow experience* was introduced in 1975 by Csikszentmihalyi from the school of positive psychology; it is intimately related to intrinsic motivation. More recently flow experience has been measured by Smyslova & Voiskounsky (2005). The major factors that contribute to flow experience are: there must be a good match between the user's needs and skills, and the inner structure of the programming tools, the quality of the live-help the system provides, and that manuals and text-books match the interests of every user. Moreover Smyslova & Voiskounsky propose that the experiential tutorial should lose descriptiveness and contain actual projects that require an increasing degree of difficulty.

Design: The importance for high-quality software engineering emerges from the contributions at this conference. This emphasis recognizes that the greatest influence of Web-based learning environments is the creativity of the designer's skills for the interface mechanisms. Throughout the analysis of these conference papers we can see the combination of the engineering domain influencing the production of the educational artefacts.

As expected, the participation from such a broad spectrum of researchers provided rich insights into the cultural factors and Web-based pedagogical determination of e-Learning effectiveness. The more important conclusions reached by the authors are summarised below:

1. Web-based learning platforms must recognise the need for *context-awareness*, particularly in the education of student cohorts from diverse cultural backgrounds.
2. Knowledge retrieval effectiveness, and more generally, knowledge management, during student interaction is vital in terms of student satisfaction with the learning environment.
3. The establishment of common standards for student learning assessment is imperative as we move towards a global learning grid, involving

credit transfer requirements and knowledge level certification. The growth of industry-based skill certification processes integrated with formal instructional models (*didactic knowledge*) needs recognition and endorsement, if changes in vocational requirements and blended learning practices are to be met.

4. The need for more effective mechanisms for transferring pedagogical practice and learning technologies between learning communities is now paramount, if the rich diversity of instructional design in Web-based learning environments is to be harnessed.

5. FUTURE WEB-BASED EDUCATIONAL RESEARCH

Future progress in Web-based instructional environments requires inspirational research, typically conducted within educational technology institutes, as discussed by Kunnath (2004). However, it is the requirement for life-long learning when competing in a global economic environment that has challenged many of the assumptions underpinning current educational practice.

Three new requirements may be identified:

1. *Learning Reinforcement* strategies are imperative, given current demographic trends in many countries, and the consequent need to re-employ older workers! In this context podcasting in its broadest sense is emerging as one of the brightest education management tools on the horizon. Although there may be a common perception this technique involves over-the-air broadcasting, podcasting lends itself in a natural sense towards education management. Many public domain Web sites are already springing up offering thousands of podcasts. The question will remain as to whether this type of education management environment can entice the practitioners to leave their more traditional class room management to reach towards audio and video feeds.
2. The cost to individuals of renewing skills and knowledge in a user-pays environment is forcing a critical appraisal of learning effectiveness. While Web-based learning environments are recognised as having a critical role to play in improving the cost-effectiveness of postgraduate training, the research challenge is to develop an appropriate methodology for operational audit of learning effectiveness.
3. The importance of the moving image is slowly being recognised in context-awareness processes, and while suitable technology platforms are still evolving, HCI using moving images can be expected to provide an important new horizon.

6. REFERENCES

Csikszentmihalyi, M. (1975). Beyond Boredom and Anxiety: The Experience of Play in Work and Games. San Francisco, Jossey-Bass.

Flicker, B. (2002). Working at Warp Speed: The New Rules for Project Success in a Sped-up World. San Francisco, Berrett-Koehler.

Gagne, R. M. (1985). The Conditions of Learning: And the theory of instruction. NY, Holt/Rinehart/Winston.

ICCE2004-Full-Proceedings (2004), E. McKay, Ed. Acquiring and Constructing Knowledge Through Human-Computer Interaction: Creating new visions for the future of learning. International Conference on Computers in Education (ICCE2004), held in Melbourne, Australia - Nov 30 - Dec 3, Common Ground Publishing: ISBN 1 86335 573 1.

Li, L., Zheng, Y., Ogata. H. & Yano, Y. (2005) Towards ubiquitous learning space, E. McKay, Ed. International Conference on Computers in Education (ICCE2004), held in Melbourne, Australia - Nov 30 - Dec 3, Common Ground Publishing: ISBN 1 86335 573 1, 1315-1325.

Preece, J. (1994). Human-Computer Interaction. Harlow - UK, Addison-Wesley.

Smith, A. (1997). Human Computer Factors: A Study of Users and Information Systems. Berkshire, McGraw-Hill.

Smyslova, O. V. & Voiskounsky, A. E. (2005), The Importance of Intrinsic Motivation in Usability Testing. 3rd International Conference on Universal Access in HCI, held at the 11th International Conference on HCI (AC, UAHCI, HIMI, OCSC, VR, U&I, EPCE): Las Vegas, HCI-International.org. Full Proceedings on CD: ISBN: 0-8058-5807-5. Viewed on 15/01/05 at http://www.hci-international.org/.

ICT PD 4 Me!

Christopher D O'Mahony
St Ignatius' College, Riverview, Sydney, Australia

Abstract: Researchers and practitioners recognise the crucial links between ICT Access and ICT Ability in promoting ICT Use. This paper is the third in a series exploring this link, and concentrates specifically on approaches to ICT Ability. This paper considers literature and research relating to ICT PD from a variety of sources and countries. Building on a model proposed at the ITEM2002 conference, the paper explores research conducted in a selection of schools in England, and then discusses one implementation of ICT PD programmes in Australia. Results of this implementation are discussed, concluding that it may be a valuable model for other schools seeking to leverage ICT use for educational innovation.

Keywords: School information systems, professional development, educational management, information technology.

1. INTRODUCTION

In the past two decades during which the phenomenon of information and communications technologies (ICT) has burst onto the educational scene, many commentators have noted the often haphazard and ad hoc nature of ICT assimilation in schools. The road from 'initial adoption' to 'embedded use' is littered with war stories of hardware obsolescence, software incompatibility, systemic underinvestment and lack of user competence (Cuban 2000, Bechervaise & Chomley 2003). It has been, and continues to be, a journey of 'trial and error' to determine what models are effective in embedding ICT into teaching, learning and administration.

This paper provides some perspectives on the use of staff professional development to embed ICT use into a school setting. It builds on previous work by the author as well as other ITEM commentators, linking theoretical models, survey and case study research, and empirical evidence. Section 2 describes an ICT Competence model initially developed in 2002. In contrast, Section 3 reviews results from a survey conducted in the UK in 2003,

Please use the following format when citing this chapter:

O'Mahony, C.D., 2007, in IFIP International Federation for Information Processing, Volume 230, Knowledge Management for Educational Innovation, eds. Tatnall, A., Okamoto, T., Visscher, A., (Boston: Springer), pp. 167–177.

highlighting the issue of ICT competence. Section 4 presents a case study from one Australian school. Empirical evidence from this is analysed in Section 5, showing the effectiveness of the PD programme. Sections 6 and 7 conclude by demonstrating that there are genuine and measurable improvements in ICT use through a robust ICT PD programme and strategy.

2. AN ICT COMPETENCES MODEL

It is generally acknowledged in the literature that getting ICT professional development 'right' is difficult to achieve (Selwood et al 2000, Visscher & Brandhorst 2001). At the same time, despite the difficulties, there is also recognition that there are critical links between staff ICT ability and use, and staff ICT ability and student ICT ability (Kennewell et al 2000, Russell et al 2000, NGfL 2002). In 2002 a discussion group was formed at the ITEM conference in Helsinki to consider core competences required for ITEM. In part, this was in response to a recognition that many ICT PD efforts had failed. A key outcome of that discussion group was a proposed competences framework (Selwood et al, 2002). This model identified 36 competences across three dimensions – knowledge domain, organisation level, and stage of growth – as described in the following sections.

2.1 The X Axis: ITEM Knowledge Domains

By deconstructing the ITEM group's basic terms, four core knowledge domains present themselves – information, technology, education and management. ITEM links these four domains inextricably and it is in their interactions that ITEM has its uniqueness, and it is important to distinguish the different competences required in each of these knowledge domains.

2.2 The Y Axis: Organisational Levels – Operational, Tactical, and Strategic

The management process is primarily concerned with decision making. Drawn from the management literature decision making can occur at three different levels operational, tactical, and strategic. Whenever for example, head teachers and governing bodies in a school form a vision of where the school is going, establish aims and objectives, prioritise them and develop a plan for the accomplishment of these objectives, they are involved in strategic decision making. Whenever they are making decisions concerned with the implementation of the school's development plan, their decisions may be considered as tactical. Finally, whenever they have to carry out clear and specific tasks they are making operational decisions. Other authors have considered these organisational levels in terms of the degree of structured thinking, whereby operational decisions are highly structured, tactical decisions are semi-structured and strategic decisions tend to be unstructured.

2.3 The Z Axis – Stages of Growth

This axis reflects the degree of sophistication of the organisation's IT effort, as seen in well-exercised 'stages of growth' theories, including Nolan (1979), Visscher (1991) and Galliers & Sutherland (1991).

Each of these theorists offered a model with certain relevance to the ITEM competence debate. Both the Nolan and Galliers & Sutherland models were perceived as too generic for the ITEM domain, whereas the Visscher model was perceived as being the most relevant. Ultimately, the stage labels used were Initiation, Expansion, and Embedded.

2.4 Advantages of the ITEM Competence Model

The proposed model offered certain advantages, as follows:

- It enables us to map existing policies and programmes, and to investigate goodness of fit of those policies and programmes;
- It enables us to map ITEM competences to existing job descriptions within educational institutions;
- It is platform-independent;
- It is descriptive, as well as prescriptive;
- It helps us in the development of more appropriate policies and programmes for ITEM competences.
- It assists in development of a post-graduate ITEM curriculum;
- It enables the certification of other courses against the criteria noted in these competences.

3. THE UK SURVEY

In late 2002, as part of a wider study, 1366 school staff responded to questions related to ICT access, ability and use. Results were analysed from Nov 2002 to Feb 2003, according to strata such as School, Region, Age, Phase (Primary, Secondary) and Department. Data were analysed to explore relationships between dependent and independent variables.

3.1 Ability with ICT

In one question, staff were asked to self-assess their current ability with a selection of applications. In a subsequent question, they were asked to nominate their desired level of ability with the same selection of applications. Tables 1 and 2 summarise responses to these questions.

Table 1. Perceived current ability (Where 0 = Non-existent, 3 = Advanced)

ITEM	Primary	Secondary	AVG
Word	2.03	2.16	2.13
Excel	1.27	1.54	1.47
Powerpoint	0.77	0.95	0.90

ITEM	Primary	Secondary	AVG
Access	0.50	0.65	0.61
Publisher	1.00	0.80	0.85
Email	1.82	2.00	1.95
Web Search	1.70	1.88	1.83
Web design	0.19	0.31	0.28
Projectors	0.25	0.44	0.39
Digital Whiteboard	0.43	0.39	0.40

Table 2. Desired future ability (Where 0 = Not Relevant, 3 = Advanced)

ITEM	Primary	Secondary	AVG
Word	2.47	2.54	2.52
Excel	2.01	2.21	2.16
Powerpoint	1.78	1.97	1.92
Access	1.28	1.43	1.39
Publisher	1.80	1.66	1.70
Email	2.27	2.38	2.35
Web Search	2.18	2.29	2.26
Web design	1.01	1.10	1.08
Projectors	0.91	1.21	1.13
Digital Whiteboard	1.42	1.48	1.47

In order to determine 'professional development priorities', the differential between "current" and "desired" was calculated for each application. These priorities were calculated for primary and secondary phases, and are shown in Table 3.

As can be seen from the table, most respondents expressed confidence in their ability with core applications such as word processing, email and internet searching. The main priorities perceived by respondents were for training in presentation-based applications (hardware and software), the clear implication being that they recognised genuine benefit for teaching and learning from these skills.

When asked to identify the factors that were preventing them from making more use of ICT, staff nominated 'Lack of Training' as a major inhibitor (O'Mahony 2004).

Table 3. ICT PD Priorities

	Sample-Wide	Primary	Secondary
Priority	**Application**	**Application**	**Application**
1	Smartboards	Powerpoint	Powerpoint
2	Powerpoint	Smartboards	Smartboards
3	Publisher	Publisher	Publisher
4	Web Design	Web Design	Data Projectors
5	Data Projectors	Microsoft Access	Microsoft Access
6	Microsoft Access	Microsoft Excel	Web Design
7	Microsoft Excel	Data Projectors	Microsoft Excel
8	Web Searching	Web Searching	Web Searching
9	Email	Email	Email
10	Microsoft Word	Microsoft Word	Microsoft Word

4. AN AUSTRALIAN CASE STUDY

4.1 Background

The case study school is an independent day and boarding college for boys in Sydney, Australia. Although initially hesitant to embrace ICT innovations in the early 1990s, the school's management realized in 1994 that a number of 'push' and 'pull' factors were at work which required a whole-school strategy for ICT. After engaging external consultants, the school tabled its first ICT Strategic Plan in late 1995. This plan made provision for an extensive rollout of fibre-optic and category5 cabling throughout the school site, an ongoing programme of investment in end-user hardware and software, a review of curriculum outcomes to incorporate ICT elements, and provision for staff training and support. It is interesting to note the extent to which ICT has become embedded in the school's culture by considering the following comparisons between 1994 and 2004 (Table 4).

In addition to a high level of ICT provision within the school, members of the school community (staff, students and parents) also exhibited high levels of access to ICT outside school. In 2003, 95% of staff reported access to internet and email from home. Students and parents reported high levels of access to computers (97%), and high levels of access to internet and email in the home (89%). These metrics are consistent with similar statistics from other developed countries around the world (Research Machines 2000, DfES 2001, National Statistics 2002, Ofsted 2002).

Table 4. Evolution of ICT in case study school

Dimension	1994	2004
Students	1100	1550
Academic Staff	120	170
Support Staff	50	80
Student Computers	60	450
Student-Computer ratio	1:19	1:3.5
Staff Computers	15	120
(Academic) Staff-Computer ratio	1:8	1:1.5
% Computers with internet access	5%	100%
% students with email accounts	0%	100%
% staff with email accounts	0%	100%
Servers	2	35
ICT Support staff	1	10
Annual ICT spend (as % of total spend)	1%	8%

4.2 ICT Management model

There has been an emerging perception throughout the school that ICT is an enabler, providing increased efficiencies and effectiveness in administration, and adding value to teaching and learning (Kennewell et al 2000, Passey 2002). In educational institutions ICT can become either a bridge or a chasm. ICT can be a bridge in that it has the potential to:

- directly support pedagogical efforts in the teaching and learning context, in terms of delivering educational and applications software to the classroom;
- directly support the backoffice functions of the school;
- indirectly minimize the impact of necessary administrative functions that teachers, students and parents are required to perform.

ICT can be a chasm in the sense that it:

- so frequently falls short of the expectations and overblown promises of suppliers, developers and purchasers;
- is often perceived as a weapon for administrative control, rather than a tool for educational empowerment.

Arising from the school's efforts to embed ICT into its operations is a growing recognition of key factors that enable ICT to flourish, and thus for the school to be more effective in its overall development (Kirkman 2000). These factors form a 6-point model for achieving confident use with ICT. Within the school, this model exists in a specific context, whereby ICT has increasingly become a fulcrum for change in the organisation's culture. In summary, the 6-point-model comprises: policy, executive commitment, ICT resources, professional development, evaluation /appraisal, student learning. Details of this 6-point model are outlined in the following sections.

4.2.1 Curriculum ICT Policy (Strategic)

The school, through its ICT Strategic Plan, makes a clear statement of intent and direction concerning use of ICT in curriculum areas. This is visible in school documentation at senior management level, and is internalised throughout the curriculum (Kennewell et al 2000). "In ... successful schools, senior management do more than provide support for the IT Coordinator's policy; rather the IT policy is viewed as emanating from senior management" (ACCAC 1999). ICT policy seeks to articulate well with the school's business and strategic development plans (Yee 2000).

4.2.2 Department Commitment (Tactical)

At the Department level, ICT policies exist which articulate with the wider ICT strategy, and provide necessary detail and context for the respective curriculum area. These policies express the department's commitment to ICT professional development, and specify expectations of ICT use in the classroom, both in terms of minimum hours and ICT-based tasks (Newton 2003, Lambert & Nolan 2003).

4.2.3 ICT Resources

A pre-requisite to success with school ICT is the provision of sufficient, reliable and up-to-date resources (ACCAC 1999). These resources include network infrastructure, workstation and peripheral hardware, software and human resources. Table 4 (see Section 4.1) indicates the school's investment in resources over a ten year period. Strong project management methodologies have been applied to ensure that the school gains good value for money, recognising that inferior products do more damage than good. Rigorous criteria are used in selecting hardware and software applications, such as ease of integration and ease of use (Stevenson 1997).

4.2.4 Teacher Professional Development

Hiring external trainers has often been the only option for schools but, increasingly, schools are considering appointment of dedicated training staff within the overall ICT function (Watson 2001), as is the case with the subject school (O'Mahony 2002). As well as having ICT resources and policies regarding use of ICT in teaching, learning and administration, the school has implemented a robust and measurable professional development programme (Donnelly 2000, Russell et al 2000, Webb & Downes 2003). This programme has six main components, as described in Section 5.

4.2.5 Staff Appraisal and Review

The appraisal and review process gives crucial feedback for all aspects of the model. To drive home the message concerning the school's commitment to ICT, effective classroom use of ICT is a performance indicator for staff. The reviewer can flag the reviewee's ICT training needs, which is communicated to the training function/coordinator, who organises/delivers the required training. Once completed, confirmation of training is passed back along the chain. As well as providing feedback on staff ability, the appraisal and review process offers the opportunity to flag any issues concerning ICT resourcing or access. These issues, too, are forwarded to the relevant person/function. Collectively, they will assist in the formation of subsequent ICT strategies (Barnes & Greer 2002, Dowling 2003).

4.2.6 Student Learning

The ultimate aim of this model, and in particular the Staff ICT PD programme, is improvement of student learning. Thus, complementary to a Staff ICT skills programme is a cross-curricular student ICT skills programme. Transcending the use of ICT in specific subjects, this provides broad-based exposure to generic ICT skills, including keyboard familiarity, word processing, spreadsheets, presentation graphics, internet searching, critical analysis of web-based data, and 'appropriate ICT use' (NGfL 2002a, NGfL 2002b, Hruskocy et al, 2000).

5. THE AUSTRALIAN ICT PD STRATEGY

The case study school, through its ICT Training Service, encourages the continuing ICT professional development of all school staff. Services are reviewed regularly to ensure their ongoing relevance to the needs of the school community, and articulate closely with the wider professional development service offered within the school. Components of the ICT PD strategy are as follows:

5.1 ICT PD components

Initial orientation – all new staff to the College are offered an introductory ICT training session as part of their orientation. This session introduces staff to the school network and the wide array of information and communication technologies that are available.

Formal Training – a comprehensive schedule of formal ICT training has been developed. As part of the school's commitment to ICT professional development, normal teaching loads include one period per cycle (fortnight) specifically set aside for ICT training. These PD periods are delivered within the normal timetable. Session topics are negotiated with faculty heads.

E-Learning – the school has purchased site licences for a number of E-learning products. These are available through the network, and can also be accessed remotely. They are self-paced, and are complementary to the formal training sessions noted above.

BITES – "Bits of IT Excellent Stuff" is a series of after-school presentations held in a larger school hall. Visiting speakers present and demonstrate innovative ways of using ICT in teaching and learning. Usually there is one presentation each Term, the evening, open to staff and parents.

Surgeries – in addition to the more formal class-based PD, school IT training staff make themselves available for informal question & answer sessions. These are generally one-on-one, but can also take the form of small-group training.

External Events – throughout the year, a number of external organisations offer a selection of practical workshops, master classes, conferences, demonstrations and exhibitions related to ICT and education. The school budgets for staff participation in these events, with an expectation that participating staff present feedback at a relevant forum.

5.2 ICT PD Survey

In 2004, a survey was conducted among staff to explore the efficacy of the ICT PD strategy. The survey comprised two questionnaires – one administered at the beginning of the school year to provide baseline data, and a second conducted at the end of the year. Both questionnaires had essentially identical questions, thus forming a valid pre-test/post-test model.

The following table shows a summary of results for both questionnaires (Table 5).

Table 5. Questionnaire results (Scale = 1 to 5)

Dimension	Feb-04	Nov-04	change	% diff
Windows	3.57	4.02	0.65	13%
Email	3.28	3.98	0.70	14%
Internet	3.18	3.70	0.52	10%
ICT in Class	2.18	2.50	0.37	7%
Word	3.33	3.71	0.38	8%
Excel	2.66	2.99	0.33	7%
Powerpoint	2.59	3.17	0.58	12%
Marks	3.53	4.11	0.58	12%
Workbench	n/a	1.96		
Training	n/a	3.73		

6. DISCUSSION OF RESULTS

Analysis of the results of the 2004 survey demonstrated a number of things. Firstly, the February responses corroborated the findings of the 2003 UK survey, showing that staff demonstrated a core competence across a range of ICT applications. Secondly, the November responses demonstrated that, through the intervention of the ICT PD strategy, significant improvements were made in ICT competence across these applications. Other observations regarding the impact of the ICT PD strategy includes the following:

- Increased ICT competence has built an expectation among staff for increased ICT access. Thus the school has recognised that increased ICT access and increased ICT competence are linked in an 'expectation spiral'.
- Increased ICT competence has led to increased enthusiasm for ICT use, and also for ICT PD itself, in recognition that it really does make a difference in teaching and learning.
- The school is now incorporating an expectation of ICT competence into its staff promotions policy. Staff must demonstrate successful completion of ICT PD components in order to be considered for internal promotion positions.
- The school's ICT support function has noted a reduction in service calls, directly related to an increase in staff resilience and self-help, brought about by the ICT PD Strategy.

7. CONCLUSION

This paper has brought together research that spans many years and many countries, all linked by the central theme of ICT professional development, and its impact on overall school effectiveness with ICT. From an essentially theoretical starting point, the paper has shown how a specific

ICT PD strategy can be devised and implemented. Unlike some other models of ICT PD that have been attempted through the years, this small-scale strategy has demonstrated genuine improvements in ICT competence, as well as positive outcomes for ICT effectiveness in teaching, learning and administration. It is anticipated that other schools may gain similar improvements by adopting a similar strategy.

8. REFERENCES

ACCAC (1999), Whole school approaches to developing ICT capability. Cardiff: ACCAC.

Barnes, A., & Greer, R., (2002) "Factors affecting successful R-12 learning communities in web-based environments", Proceedings: ACEC2002, Australian Computers in Education Conference, July 2002.

Bechervaise, N.E., & Chomley, P.M.M., (2003), "E-lusive learning: innovation, forced change and reflexivity", Proceedings: E-Learning Conference on Design and Development, Melbourne: RMIT, November 2003.

Cuban, L. (2000), Oversold and Underused: Computers in the Classroom. Cambridge, MA: Harvard University Press.

DfES (2001), ICT Access and Use: Report on the Benchmark Survey, DfES Research Report No 252. London: Department for Education and Skills.

Donnelly, J. (2000). Information Management Strategy for Schools and Local Education Authorities – Report on Training Needs. http://dfes.gov.uk/ims/JDReportfinal.rtf . DfES, London

Dowling, C., (2003), "The role of the human teacher in learning environments of the future", Proceedings: IFIP Working Groups 3.1 and 3.3 Working Conference: ICT and the Teacher of the Future, Melbourne, 2003.

Galliers, R.G. & Sutherland A.R., (1991), Information systems management and strategy formulation: the 'stages of growth' model revisited, Journal of Information Systems, 1, 1991.

Friedlander, J., (2004), "Cool to be wired for school", Sydney Morning Herald, April 16, 2004.

Hruskocy, C., Cennamo, K.S., Ertmer, P.A., Johnson, T., (2000) "Creating a community of technology users: students become technology experts for teachers and peers", Journal of Technology and Teacher Education, Vol 8, pp69-84.

Kennewell, S., Parkinson, J., and Tanner, H., (2000), Developing the ICT-capable School. London: Routledge Falmer.

Kennewell, S., (2003), "Developing research models for ICT-based pedagogy", Proceedings: IFIP Working Groups 3.1 and 3.3 Working Conference: ICT and the Teacher of the Future, Melbourne, 2003.

Lambert, M.J., & Nolan, C.J.P., (2003). Managing learning environments in schools: developing ICT capable teachers. In Management of Education in The Information Age - The Role of ICT. Edited by Selwood I, Fung A, O'Mahony C. Kluwer for IFIP. London

National Statistics (2002). www.dfes.gov.uk/statistics/db/sbu/b0360/sb07-2002.pdf

Newton, L., (2003). Management and the use of ICT in subject teaching – integration for learning. In Management of Education in The Information Age - The Role of ICT. Edited by Selwood I, Fung A, O'Mahony C. Kluwer for IFIP. London

Nolan, R.L., (1979), Managing the Crises in Data Processing, Harvard Business Review, 57, 2, March 1979, pp 115-126.

Ofsted (2002). ICT in Schools, Effect of Government Initiatives. http://www.ofsted.gov.uk /public/docs01/ictreport.pdf. DfES , London.

O'Mahony, C.D. (2002), Managing ICT Access and Training for Educators: A Case Study, Proceedings: Information Technology for Educational Management (ITEM2002 conference), Helsinki.

O'Mahony, C.D. (2004), E-Learning component evolution and integration: a case study, Proceedings: International Conference on Computers in Education (ICCE2004), RMIT, Melbourne.

Research Machines PLC (2000) The RM G7 (8) Report 2000 comparing ICT provision in Schools, Abingdon: RMplc.

Schiller, J., (2002), "Interventions by school leaders in effective implementation of information and communications technology: perceptions of Australian Principals", Journal of Information Technology for Teacher Education (JITTE), 11, 3, 2003.

Selwood, I. (1995). The Development of ITEM in England and Wales in Information Technology in Educational Management. Edited by Ben Zion Barta, Moshe Telem and Yaffa Gev. Chapman Hall for IFIP, London, UK.

Selwood, I.D. & Drenoyianni, H. (1997). Administration, Management and IT in Education in Information Technology in Educational Management for the Schools of the Future. Edited by Fung A, Visscher A, Barta B and Teather D. Chapman & Hall for IFIP. London, UK.

Stevenson, R., (1997), Information and Communications Technology in UK Schools: an independent inquiry (The Stevenson Report).

Visscher, A.J. & Brandhorst, E.M. (2001). How should School Managers be Trained for Managerial School Information System Usage? In Pathways to Institutional Improvement with Information Technology in Educational Management. Edited by Nolan, C.J.P., Fung, A.C.W., & Brown, M.A. Kluwer for IFIP. London

Visscher, A.J., (1991), School administrative computing: a framework for analysis, Journal of Research on Computing in Education, 24, 1, Fall 1991, pp 1-19.

Watson, G (2001), "Models of information technology teacher professional development that engage with teachers' hearts and minds", Journal of IT for Teacher Education (JITTE), 10, 1-2, 2001.

Webb, I., & Downes, T., (2003), "Raising the standards: ICT and the teacher of the future", Proceedings: IFIP Working Groups 3.1 and 3.3 Working Conference: ICT and the Teacher of the Future, Melbourne, 2003.

Yee, D.L., (2000), "Images of school principals' information and communications technology leadership", Journal of IT for Teacher Education (JITTE), V 9, No 3, 2000.

ITEM in Botswana and Uganda

R. Bisaso, O. Kereteletswe, I.D. Selwood, A.J. Visscher
Department of Higher Education, Makerere University. MoE, Botswana. School of Education, University of Birmingham. Faculty of Behavioural Sciences, University of Twente

Abstract: This article reports on the lessons learnt from the implementation of a computerized information system (CIS) for managing human resources at the Ministry of Education in Botswana, and on the usage of CISs in the management of secondary schools in Uganda. The findings from these African studies portray the levels of usage, their impact and the critical success factors that most influence the utilization of the CISs. In both countries, clerical usage of the CISs was reported. User training is reported as the most important determinant of CIS usage in both Uganda and Botswana. In Uganda, managerial usage by school managers is very limited, but users are generally positive concerning the effects of CISs use. In Botswana, the direct usage by managers is also limited as is use for decision-making. It is concluded that wider and better CIS usage can be promoted by carefully designed user training, grounded on a thorough analysis of the needs of the user group.

Keywords: Developing countries, management information systems, educational management

1. INTRODUCTION

Over the last two decades, the use of management information systems for education has grown enormously in the industrialized countries and now developing countries are starting to utilize the potential of these systems. Administrative reforms in Botswana geared to improving efficiency in the public sector together with a rapidly expanding education systems are the foundation upon which the computerization of the activities of the Ministry of Education (MOE) were grounded. The aim of the introduction of a computerized personnel management systems was provision of adequate and accurate staff information for management reporting and planning purposes. In Uganda, since the inception of universal primary education (UPE) in 1997, enrolments in secondary schools have increased greatly as a result of the escalating primary school leavers. For example, in 1999, Uganda had

Please use the following format when citing this chapter:

Bisaso, R., Kereteletswe, O., Selwood, I.D. and Visscher, A.J., 2007, in IFIP International Federation for Information Processing, Volume 230, Knowledge Management for Educational Innovation, eds. Tatnall, A., Okamoto, T., Visscher, A., (Boston: Springer), pp. 179–186.

only 625 registered secondary schools but by 2001, the number of registered secondary schools had risen to 1850 (Aguti, 2002). Owing to this exponential effect of UPE, secondary schools have now started utilizing the potential that computerised information systems (CISs) offer. In this paper we report on studies carried out in Uganda and Botswana respectively, that provide insight into how these two developing countries benefit from the potential of ICT for administering and managing education, which problems they meet, and which are the critical success factors for implementing CISs.

2. RESEARCH QUESTIONS AND RESEARCH FRAMEWORK

The two studies were guided by the following research questions, which were formulated with regard to the variable clusters in the Visscher model (1996; 2001, see Figure 1).

1. To what extent and how are the computerized information systems used in secondary schools and in the Ministry of Education in Uganda and Botswana respectively?
2. What are the positive/negative effects of the use of the computerized information systems on managerial activities in secondary schools and in the Ministry of Education?
3. What problems are faced in the introduction and utilization of computerized information systems in these two developing countries?
4. Which factors influence the utilization of the computerized information systems in these two developing countries?

Figure 1 reflects the assumption that the extent of usage of a CIS depends on the influence of CIS quality, implementation process features and organizational features. Therefore, it is expected that the higher the perceived CIS quality, the more the implementation process promotes CIS usage, and the stronger the correspondence between CIS features and the nature of the organization, the more intense CIS use (block D) will be. The extent and nature of CIS use may yield both positive and negative effects (block E). While there has been extensive application of the Visscher model in research in the developed countries (Visscher 1996; 2001), its application in developing countries has been exploited for the first time in the studies that we report on here.

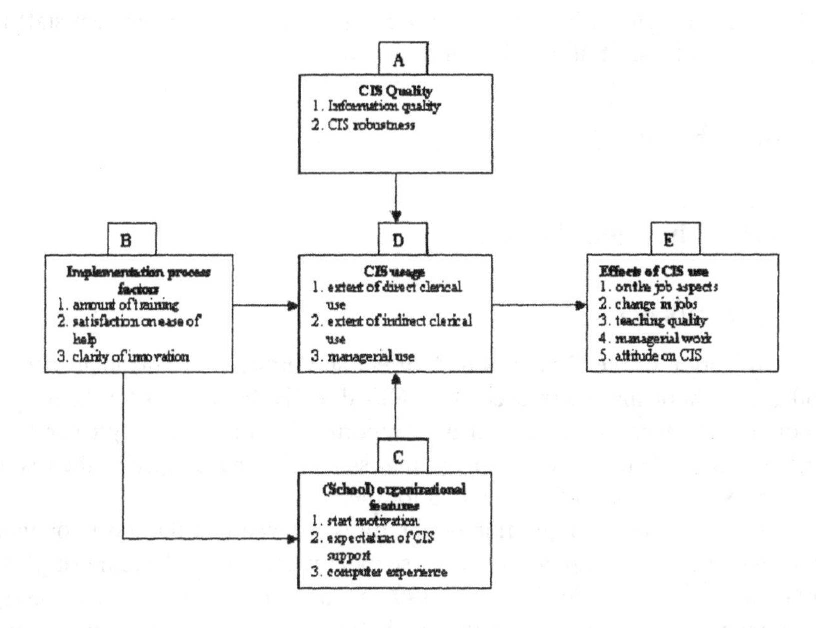

Figure 1: Factors influencing the usage and impact of CIS in developing countries (Visscher1996 and 2001)

3. METHOD AND DATA ANALYSIS

The study in Uganda was carried out in the four school districts that have the biggest and most well developed schools. The implementation of Information Technology and Educational Management (ITEM) is recent and therefore it was felt best to evaluate schools where the probability of ITEM's occurrence would be greatest. Fifty-five schools were selected to participate in the study using a non-probability sampling method. Respondents were school management personnel (one per school). The user response rate was 100% after follow-up activities. Thirty-four schools reported that they were CIS-users and the other 21 schools indicated that they either had an information system but were reluctant to use it, or they were in the advanced stages of acquiring one.

In Botswana, 168 questionnaires were distributed to the departments involved in the management of teachers at the Ministry of Education. These were Teaching Service Management, Secondary Education, Primary Education, and Regional Education Offices. The response rate was 76%. Respondents included clerks, middle managers and senior managers.

Data analysis involved the computation of frequencies and carrying out regression analyses, the latter to establish those factors that predict the level of direct and indirect information system use by school management staff in Uganda and by staff of the MOE in Botswana.

4. RESULTS

4.1 The Ugandan study

4.1.1 CIS use

Utilization of the CISs was both direct and indirect. In Ugandan schools 60% of school managers used their CIS directly for one to ten hours per month. However, 66% of managers reported indirect use of greater than eleven hours. Thus it would appear that schools managers prefer their staff to retrieve information from the CIS for them.

The analysis of the questionnaire returns showed that the most common CIS modules in Ugandan secondary schools were financial monitoring and planning (91%), student records (90%) and student assessment (83%). Commencing CIS implementation with financial planning and monitoring and pupil records was also common in developed countries (Selwood 1995), and this is probably due to managers being concerned regarding their liability if financial records are found to be inaccurate. Furthermore, school funding generally relates to pupil numbers, and age of pupils, therefore pupil data is needed by the finance module (and for future implementation of most other modules). The least available modules were library management and student attendance.

4.1.2 Effects of CIS use

Over half the respondents were positive about the effects of CIS usage. The highest level of response concerned 'evaluation of school performance' with 81% feeling this had improved. Interestingly, this effect 'evaluation of school performance' did not record any negative response. 'Utilization of school resources' had 74% of staff reporting improvement and 'information for curriculum planning' was reported improved by 69% of respondents. 'Stress' and 'workload' are reported to have improved according to 66% and 69% of the respondents respectively, thus staff feel less stressed and that their workload has reduced. Only 'internal communication with colleagues' registered a significant negative percentage (23%). However even for this category, 55% were positive that there were benefits.

4.1.3 Problems

Even though positive effects of implementing the schools' CISs were reported in the previous section, the quality of the CIS failed to meet the expectations of many users with 62% of the respondents noting that the systems did not fully provide the information they needed. Furthermore, the expectations of the systems exceeded the most optimistic projections of them e.g. 69% of the respondents felt the quality of teaching had been influenced by the CIS. This remains to be empirically proven especially without any established relationship between the schools' clerical work and the teaching and learning process. Lack of prior experience in using computers either at home or work was very evident from the respondents but this is perhaps a symptom of the lack of use of information technology in developing countries.

4.1.4 Critical Success factors

The results of regression analysis of the data gathered in the Ugandan study demonstrates clearly that the "amount of external training" is a factor predicting system usage in secondary schools. 'Support from the system administrator' also plays an important role possibly due to the lack of experience of school managers, since information system utilization is a recent innovation in Uganda. Together both factors explain 35% of the variance in CIS use. Furthermore the analysis demonstrated if the 'amount of external training' increases by 1 standard deviation then CIS-use increases by .39 of a standard deviation. Interestingly, if 'support from system administrator' increases by 1 standard deviation CIS usage increases even more, by .61 of a standard deviation.

4.2 The Botswana study

4.2.1 CIS use

The analyse of questionnaire returns clearly showed that direct use of the CIS was not high in Botswana with 34% of respondents not using it directly and 59% using it directly for between 1 to 4 hours per week. Indirect use appears somewhat better with only 14% reporting no use and 73% reporting use between 1 to 4 hours per week. Further analysis of the data by type of user revealed that managers tended to be indirect users and clerical staff were more likely to be direct users (Kereteletswe and Selwood, 2004).

4.2.2 Effects of CIS use

Even though as detailed above CIS usage was not as high as would be expected the data gathered clearly demonstrates positive attitudes with respect to the effects of CIS. Combining better and much better only 'processing gratuity' and 'processing terminal benefits' receive less than 50% positive response. The highest level of improvement was seen for 'management of teacher transfer' and 'errors in posting' both achieving 94%. Interestingly there appears to be a contradiction in the results when we compare 'time needed to carry out duties' (88% reporting an improvement) with 'work load' (56% reporting an improvement). The reasons for this may be either that with the introduction of CIS some staff have had other duties added to their workload, or as Riggs (1964) asserts, in developing countries (prismatic societies) there is a tendency, when innovations are introduced, for the old and new to sit side by side. That is to say, the old system continues and runs in parallel with new, thus duplicating work, and increasing workload.

4.2.3 Implementation constraints in Botswana

Based on their experience of implementation at the Ministry, the respondents were asked to rank the potential implementation constraints likely to be faced by the MoE in rolling out the CIS to Regional Education Offices (REOs) and schools. The factors were ranked in order of priority with 1 (the most important) and 8 (the least important). The results of the Friedman test show that 'lack of technical support' and 'lack of resources' ranked 2.10 and 2.21 respectively. This suggests that the respondents viewed these two factors as highly probable problematic factors to be faced by the MoE in rolling out the CIS to REOs and schools.

4.2.4 Critical success factors

In Botswana, analysis was undertaken on factors that most determine the level of direct or indirect use. The variables age, qualification, and gender of respondents were not significantly explanatory of differences in extent of information system use. On the other hand, the division the respondent belonged to and the position of responsibility the respondent held, explained together both direct and indirect usage at a rate of 56% and 41%.

Regression analysis showed that variance in the extent of information system use in Botswana was explained by the factors 'information quality' and 'amount of external training' for both direct and indirect use.

5. CONCLUSION

In Uganda, it is apparent that considerable progress has been made in implementing CIS in schools. However to optimise use it is equally apparent from the evidence presented here, that investment in external training for users is needed, and that as Systems Administrators are highly valued these need to be retained and supported. There are also some concerns regarding the quality of the CISs in use as respondents felt they did not fully meet their needs. However, respondents are generally optimistic and positive concerning the benefits of CIS usage in their schools.

In Botswana, after five years of system implementation, system utilization still appears somewhat ad hoc. Again external training and system quality seem to be factors that need to be addressed. Noteworthy, the factor training has been found significant in studies in the Netherlands, Hong Kong and the United Kingdom (Visscher and Bloemen, 1999; Visscher, Wild and Smith, 2003). In Botswana, it is apparent that system use has a relationship with the other cluster blocks in the Visscher model.

It is paramount that secondary schools and education ministries in developing countries like Uganda and Botswana design effective training programmes that will step up system usage. Visscher and Bloemen (2001) recommend that management-oriented training courses purposely developed for MIS usage ought to be evolved and tested to equip managers with skills necessary to bolster decision-making processes and the functioning of their institutions. Furthermore, Selwood and O'Mahony (2003) have developed a framework for ITEM competence that can be applied to develop training courses for various types of ITEM users.

Other factors that influence usage, and type of systems usage may well be cultural. The work of Riggs (1964) has already been noted in discussing system usage in Botswana and his theory may well account for some resistance to change or apparent inertia when change is implemented in developing countries. Furthermore, managers' attitudes, and work patterns may also influence system usage. Results of the analysis showed that senior management are more indirect than direct users of the CIS in Botswana, while low echelon positions use the system more directly. The literature on education management points to possible reasons why this may be the case. For example, Mintzberg (1989) asserts that managers prefer taking decisions rather than not taking them because at least something is done. Management literature, however suggests that many managers are not reflective planners and rely more on instant information received in informal ways (Visscher 1996). However, senior managers need readily available information on which to base their decisions (Visscher and Fung 2001). Thus, the lower echelon positions become information providers to the senior management.

Visscher (1996) reports that schools vary in their policy-making capacity and that the areas in which schools develop their policies often differ in degree. For example, a vibrant school policy could be developed over school resources like finance, buildings and other infrastructure while instructional matters like teaching methods, or even instructional content decisions are left to the discretion of the teachers. It is this divergence in policy-making capacity that will strongly impact on the implementation of information systems in school policy making. It is self evident that, the introduction of an information system in an area where a school has a limited policy will not improve policy-making capacity in that area, but instead precipitate additional problems associated with innovations without readiness. It is advisable that policy-making be first nurtured through organizational development prior to introduction of CIS. The usage and impact of a CIS can be extensive if schools developed a policy before installation of the CIS. It is also important that close scrutiny is made in relation to the outstanding organizational differences between schools.

To conclude, developing countries have now realized the potential that ITEM offers to the efficient and effective handling of information at both ministerial and school levels. Governments must understand that the change process is gradual and takes time; the implication being that investment in computerised information system usage should involve enough time for effective assimilation. Also, that investment in user training is essential if returns on investments in time, infrastructure and hardware are to be rewarded by effective use of CIS for Educational Management.

6. REFERENCES

Aguti, J.N. (2002). *Facing up the challenge of UPE in Uganda through Distance Teacher Education programmes.* Paper presented at Pan Commonwealth Forum on Open Learning; Transforming Education for development. Durban, South Africa.

Kereteletswe O.C., and Selwood I.D. (2005). Evaluation of the Implementation of Information Technology in Education Management (ITEM) in Botswana in Tatnall, A., Osario, J. and Visscher, A. (Eds.) *Information Technology and Educational Management in the Knowledge Society.* New York: Springer.

Mintzberg (1989). *Mintzberg on Management.* New York: Free Press.

Riggs Fred, W. (1964) *Administration in Developing Countries: The Theory of Prismatic Society.* Boston: Houghton

Selwood I. and O'Mahony C. (2003) Core Competences for ITEM – a model in Selwood, I.D., Fung, A.C.W., & O'Mahony C. (Eds.) *Management of Education in the Information Age - The Role of ICT.* pp.195-201. Boston: Kluwer.

Selwood, I.D., (1995). The Development of ITEM in England and Wales. In Barta, B.Z., Telem, M., and Gev, Y. (Eds.), Information *Technology in Educational Management* pp.85-92. London: Chapman Hall for IFIP.

Visscher, A.J. (1996). The implications of how school staff handle information for the usage of school information systems. *International Journal of Educational Research,* 25 (4), 323-334.

Visscher, A.J. (2001). Computer-Assisted School Information Systems: the concepts, intended benefits, and stages of development. In Visscher, A.J., Wild, P. and Fung, A.C.W. (Eds.) (2001). *Information Technology in Educational Management: Synthesis of Experience, Research and Future Perspectives on Computer-Assisted School Information Systems* (pp. 3-18). Dordrecht: Kluwer.

Visscher, A.J. and Bloemen, P.P.M. (1999). Evaluation and use of computer-assisted management information systems in Dutch schools. *Journal of Research on Computing in Education,* 32(1), 172-188.

Visscher, A.J. and Bloemen, P.P.M. (2001). CSIS Usage in School Management: A Comparison of Good and Bad Practice Schools. In P. Nolan, A. Fung, and M. Brown (Eds.), *Institutional improvement through information technology in educational management* (pp. 87-97). London: Kluwer.

Visscher, A.J. and Fung A.C.W (2001). Imperatives for successful implementation of school information systems. In Visscher, A.J., Wild, P. and Fung, A.C.W. (Eds.) (2001). *Information Technology in Educational Management: Synthesis of Experience, Research and Future Perspectives on Computer-Assisted School Information Systems* (pp.115-133). Dordrecht: Kluwer.

Visscher, A.J., Wild, P., and Smith, D. (2003). The Results of Implementing SIMS in English Secondary schools. In Selwood, I.D, Fung, A.C.W., and O'Mahony C.D (Eds.)(2003). *Management of Education in the Information Age. The Role of ICT* (pp. 34-44). London: Chapman and Hall.

CSCL-Based Pre-Service Teacher Program as Knowledge Building

Jun Oshima, & Ritsuko Oshima
Shizuoka University, Japan

Abstract: The study reports two design experiments on the pre-service teacher program to advance their understanding of learning as knowledge building. We designed the course with a CSCL tool. In the first year, we did not have information on students' characteristics. Analyses of students' final essays and their discourse activities on the CSCL showed: (1) that we failed to improve students' understanding at our expected level, (2) that collaborative students reached a deeper understanding than isolated students, and (3) that students' beliefs of didactic instruction resisted the new perspective on learning we introduced. In the second year, we designed the course to overcome students' resistance and to facilitate more frequent collaboration among them by: (1) making the course project-based, (2) having students in a small group use a computer for their collaboration between groups, and (3) involving them in collaborative problem-solving as learners. Results in the second year, compared with those in the first year, showed a crucial improvement of students' conceptual understanding.

Keywords: CSCL, pre-service teacher program, knowledge building.

1. DESIGN EXPERIMENTS IN TEACHER PROFESSIONAL DEVELOPMENT

The purpose of the course we designed was to have pre-service teachers acquire conceptual understanding of learning as knowledge building and how to design the classroom environment to facilitate learners' knowledge building using a CSCL technology. Since the course was scheduled as an intensive summer workshop in four to five days, we had to consider a different approach from designing similar courses in a semester (e.g., Oshima, & Oshima, 2002). Students (49 in the first year, and 51 in the second year) in the Faculty of Humanities and Social Sciences took the course as part of their requirement for teacher certificates.

Please use the following format when citing this chapter:

Oshima, J. and Oshima, R., 2007, in IFIP International Federation for Information Processing, Volume 230, Knowledge Management for Educational Innovation, eds. Tatnall, A., Okamoto, T., Visscher, A., (Boston: Springer), pp. 187–194.

In designing the course, we referred to the framework of the community of learners by Brown and Campione (1996). The curriculum was mostly students' discourse-centered. Learning contents were designed to facilitate problem-based and project-based learning with authentic tasks. For students to share their ideas, we implemented a CSCL system, Knowledge Forum®. In the first year, we took a general approach to designing the course with no information on students' characteristics. In the second year, based on our evaluation on the first year's design, we could design the course more specifically for the target students.

2. DESIGN EXPERIMENT-1

2.1 Course Design

The course was designed from the perspectives of: (1) learning contents, (2) learning activities, (3) discourse structure, and (4) the CSCL support.

First, the main goal of the course was to improve students' ability to use the theory of "distributed human intelligence" in considering lesson plans in their major subjects. To this end, we designed the contents of "distributed intelligence," and "situated learning" with examples. We also prepared contents of educational practices based on the conception of "cognitive apprenticeship (Collins, Brown, & Newman, 1989)" so that they could link their conceptual understanding to practice.

Second, students' activities were designed as repeated sequences of a lecture-based study, problem-based learning at individual, small group, or class-as-a-whole level, and reflection at a benchmark session. The course was conducted in four days. During the first two and a half days when students studied conceptual understanding of distributed intelligence and learning, each individual student used Knowledge Forum® to collaboratively construct their shared ideas on distributed intelligence and learning as participation in authentic practices. Then, during the remaining day and a half, they conducted project-based learning to study and evaluate Japanese practices.

Third, we designed a course where students engaged in three different phases of discourse: individual-based, group-based, and class-as-a-whole exchange of ideas. In the first stage, we designed students' discourse activities as individual-based exchange of ideas followed by class-as-a-whole discourse. In the second stage, we designed their discourse activities as group-based exchange of ideas followed by the benchmark session.

Fourth, we used Knowledge Forum® to facilitate students' discourse as individual-based and group-based exchange of ideas. We expected that the asynchronous discourse tool could guarantee individual contributions to the online discussion.

2.2 Evaluation

The course design was evaluated from two perspectives. First, we evaluated scores of final essays. The essay required students to design lesson plans in their major subjects. Students were allowed to write essays either individually or in collaboration. Two experts independently evaluated each essay from the four perspectives with 5 point-scales: (1) whether knowledge sharing through collaborative learning is appropriately designed, (2) whether a CSCL technology is implemented in appropriate contexts of learning, (3) whether designed activities are practically feasible in the scheduled timeline, and (4) whether goal structures are articulate enough for students to engage in their project-based learning.

Another set of measures was students' discourse activities on Knowledge Forum®. We evaluated how and what knowledge building discourse students engaged in. The analyses were conducted quantitatively and qualitatively by referring to the concept of "collective cognitive responsibility" (Scardamalia, 2002). The structure of students' discourse was analyzed with respect to "knowledge access," and "knowledge exchange." Knowledge access would happen when students read others' notes. Knowledge exchange would happen when students put their ideas on others' notes, i.e., producing note threads. Further, qualities of students' discourse were evaluated based on "four commitments to progressive discourse as science" (Bereiter, 1994). Knowledge building discourse should be scientific and progressive, and satisfy: (1) commitment to work toward common understanding satisfactory to all, (2) commitment to frame questions and propositions in ways that enable evidence to be brought to bear on them, (3) commitment to expand the body of collectively valid propositions, and (4) commitment to allow any belief to be subjected to criticism if it will advance the discourse. Two experts independently evaluated discourse in non-threaded notes or threads.

Students' final essay. Nineteen of forty-nine wrote their essays individually and 30 worked collaboratively. Because the correlation on scores between the experts was statistically significant, the sum of the four average scores was used as the essay score. The mean score of students' essays was 5.83 (SD = 2.89). A t-test on the mean scores of individual and collaborative essays showed that the mean score of collaborative essays, 6.68 (SD = 0.99), was significantly higher than that of individual essays, 4.47 (SD = 4.18), $t(47) = 2.79, p < .01$.

Results were unlike what we had been expected. We did not expect that many students (around 40% in this case) would submit their essays individually. Furthermore, the mean score of their essays was lower than our expectation. They only scored 5.83 of 16. One remarkable result in the analyses, however, was that the score of collaborative essays was higher than that of individual essays.

Students' discourse activities on Knowledge Forum®. Knowledge exchange was analyzed by comparing numbers of users engaged in producing each note thread. In the individual phase, threads were produced by 3.16 students on the average (SD = 2.00), whereas by only 1.14 groups (SD = 0.36) in the project phase. The mean note score on Knowledge Forum® was 6.90 (SD = 4.50) of 20. A *t*-test showed that the note score in the project phase, 9.54 (SD = 5.64), was significantly higher than that in the individual phase, 6.14 (SD = 3.81), $t(231) = 5.04$, $p < .01$. Thus, results suggest that: (1) students were *not frequently* engaged in knowledge exchange between groups in the project phase, (2) the quality of students' discourse was not satisfactory, and (3) students were, however, involved in more highly scored discourse in the project phase compared with the individual phase.

2.3 Discussion

Although collaborative activities like project-based learning and collaborative writing of essays were found to have positive effects on students' conceptual understanding, results showed that our design in the first year was not appropriate for facilitating such collaborative activities. We considered the following reasons for our unsuccessful course design. First, from our observation and interviews with some students after the course, we found that many students had strong resistance to the use of computers in education. Their resistance was amplified through their use of Knowledge Forum® as a main tool for communication at the individual and group-based exchange of ideas. We intended to design the use of Knowledge Forum® as a reflection tool guaranteeing each student's contribution to the online discourse. Students did not see the tool in that way. In their online discourse, they raised criticisms to educational computing with reasons such as "The use of computers makes kids' vision worse", "I think that computer-communication is not humane", and so on. The second author (the course instructor) and the first author attempted to adjust their discourse in a more productive way by proposing ideas on the effectiveness of computers in education. We failed to have them recognize the productive aspect of educational computing based on the theory of distributed intelligence.

Second, strong dependence on the asynchronous communication tool, particularly at the individual-based exchange of ideas, made learners who were unfamiliar with the style of communication hesitate to collaborate with others. Unfamiliar students frequently expressed their ideas as single notes, but most of the notes were followed up by others.

Finally, although we found collaborative activities led students to higher levels of understanding than did individual activities, their collaboration was localized within their groups but did not frequently happen beyond the group boundaries, i.e., the discourse was asymmetric between groups. One possibility might be that students had to devote their efforts to their group work. They were divided into project groups in the late stage of the course, and had to manage their groupwork. This might keep them from going beyond their group boundaries. Another possibility might be that the project-based learning was not designed in a way for students to collaborate with other groups. In the project-based learning phase, students discussed educational practices they found on the WWW. They were interested in practices in their major subjects but not others. The group structure of students with the same major subjects might not support collaboration between groups majoring in different subjects.

3. DESIGN EXPERIMENT 2

3.1 Course Design

Learning contents were prepared for having students engage in discourse on the same conceptual artifacts. What students were expected to do in the first year was to manage their own experiences in schooling, the new concepts of distributed intelligence and learning, and some examples of good practices as resources. We found that resources provided by the instructor were not sufficiently understood and appropriately utilized by students. In the second year, we attempted to design project-based learning where students themselves engage in collaborative problem solving on a task so that they could share the same experiences as learners. For this purpose, we used a task called "rescue at the Boone's meadow" from the Jasper project (CTGV, 1997). We showed students the video then asked them to think of their solutions in their groups. We used their solutions as conceptual artifacts for their discourse on Knowledge Forum® to construct their class solution. We further asked students to collaboratively report their reflections on their problem solving processes for their further knowledge building discourse on collaborative learning as a way to facilitate distributed intelligence among learners. The goal of their final essays was also changed from designing lesson plans in their major subjects to designing plans in time for integrated studies where learners have to integrate their knowledge resources from various subject matters.

Students engaged in project-based learning in all stages of the course. Students participated in project-based learning in homogeneous groups, i.e., groups of students who were majoring the same subjects, in the first stage, then worked in heterogeneous groups, i.e., groups of students who were majoring in different subjects. Projects in the first stage were designed in such a way that they collaboratively worked on problem solving for understanding new concepts of distributed intelligence and learning based on their experiences of solving the Jasper problem as learners. Projects in the second stage, on the other hand, were designed for them to consider educational practice research from KIE and WISE research group (Linn, and Hsi, 2000) and our research group on Knowledge Forum® (Oshima, Oshima, Murayama, Inagaki, Takenaka, Nagato, Yamamoto, Nakayama, & Yamaguchi, 2002). Regrouping in the second stage was legitimate for students because they needed to work with others who had different subject matter knowledge to consider lesson plans for the integrated studies. It was also legitimate for us as designers to make their group works more constructive. We had five teaching assistants who observed and videotaped students' activities in each group (one for two or three groups). The assistants informed us who had strong leadership, collaborative characteristics, and so on. With the provided information, we could arrange heterogeneous groups to make their social dynamics more constructive and collaborative.

Face-to-face discussion within groups as objects for discourse on Knowledge Forum®. In the second year, we encouraged students' discussion within groups face-to-face. We prepared computers in the room so that each group of students could use one computer collaboratively. Desks were mobile for them to create small islands for collaborative works. We also prepared another computer room near the main room just in case they wanted to use computers individually. They were instructed to report their groups' ideas, then read other groups' ideas to comment on. Their discourse on line was further used by the instructor to reflect on what they did on the day as a community and discuss with them what to do on the next day. The instructor and teaching assistants regularly elicited students' reflection on what they were doing by asking provoking questions such as "How did you understand the task the instructor told?", "Is your groupwork going well?", "If not, what do you think are problems you have to overcome?", and so on.

3.2 Evaluation

The main purpose of the analyses here is to confirm that the course design in the second year led students to better qualities of knowledge building discourse and conceptual understanding.

Comparison of students' characteristics. At the beginning of the course in both years, we conducted a multiple-choice questionnaire to ask students about subjects they were majoring in, computer literacy level (e.g., years of experiences and how often they use computers), and their typing skills. We found no significant differences by Chi-square tests.

Comparison of scores on final essays. Multiple t-tests showed that the second year score was significantly higher than the first year grand mean score, $t(98) = 8.11$, $p < .01$, and the score of collaborative essays, $t(79) = 7.40$, $p < .01$ (Table 1).

Comparisons of students' discourse activities. We compared knowledge access and exchange in the first year's project phase and the second year (Table 2). Multiple t-tests showed that significantly more groups in the second year accessed thread notes, $t(66) = 10.16$, $p < .01$, and contributed to knowledge exchange, $t(66) = 9.55$, $p < .01$.

Table 3 shows mean scores of discourse. Multiple t-tests showed that the discourse score in the second year was significantly higher than the first year grand mean, $t(293) = 12.50, p < .01$, and mean in the project phase, $t(112) = 5.95, p < .01$.

Table 1. *Mean Scores of Students' Final Essays (SDs).*

1st Year Grand Mean	1st Year Mean of Collaborative Essays	2nd Year Mean
5.83 (2.89)	6.68 (0.99)	10.18 (2.47)

Table 2. *Mean Numbers (SDs) of Groups Contributing to Knowledge Exchange and Accessing.*

	Number of contributing groups	Number of accessing groups
1st year	1.14 (0.36)	5.54 (1.60)
2nd year	5.18 (2.32)	9.48 (1.80)

Table 3. *Mean scores of discourse in notes (SDs).*

1st year grand mean	1st year mean in the project phase	2nd year mean
6.90 (4.50)	9.54 (5.64)	14.44 (2.92)

3.3 Discussion

First, the comparison of final essay scores suggests that we succeeded in designing the course better in the second year. The improvement could not be interpreted only by our instruction to students to collaboratively submit their final essay. The task was changed from planning lessons in their major subjects to integrated studies. The groups were heterogeneous in the second year. Learning activities were also revised so that they were more engaged in project-based learning. All the changes in the course design would affect better outcomes.

Second, it was found that students' discourse activities were meaningfully improved in comparison with those in the first year. The quantitative and qualitative comparisons suggest that students' discourse activities were more symmetric, i.e., they were involved in knowledge access and exchange between groups. Further, the quality of their discourse in notes was significantly higher. We consider that this would happen because students could work on their discourse as groups. The face-to-face discourse within groups scaffolded by the provoking questions could generate conceptual artifacts sharable with other groups.

Another important reason may be that they produced their sharable experiences as learners during the course work. In the first year, students had to manage how to coordinate their philosophical views on education and new concepts of distributed intelligence and learning. It was found to be very difficult to share ideas and collaboratively improve their knowledge with new concepts as conceptual artifacts when their belief systems did not accept the concepts. In the second year, they worked on the new concepts by engaging in discourse on their shared experiences constructed during the course. The discourse in the second year was more sharable among students, open to criticisms, and improvable through synthesis of different points of views. Thus, we could make students execute a different and more constructive type of "epistemic agency" (Scardamalia, 2002).

4. REFERENCES

Bereiter, C. (1994). Implication of postmodernism for science education: A critique. *Educational Psychologist, 29*(1), 3-12.

Bereiter, C. (2002). *Education and mind in the knowledge age*. Mahwah, NJ: Lawrence Erlbaum.

Brown, A. L., & Campione, J. C. (1996). Psychological theory and the design of innovative learning environments: On procedures, principles, and systems. In L. Shauble & R. Glaser (Eds.), *Innovations in Learning: New Environments for Education* (pp. 289-325). Mahwah, NJ: Lawrence Erlbaum.

Collins, A., Brown, J. S., & Newman, S. E. (1989). Cognitive apprenticeship: Teaching the crafts of reading, writing, and mathematics. In L. B. Resnick (Ed.), *Knowing, learning, and instruction: Essays in honor of Robert Glaser* (pp. 453-494).Mahwah, NJ: Lawrence Erlbaum.

CTGV (1997). *The Jasper Project*. Mahwah, NJ: Lawrence Erlbaum.

Linn, M., & Hsi, S. (2000). *Computers, teachers, peers: Science learning partners*. Mahwah, NJ: Lawrence Erlbaum.

Oshima, J., Oshima, R., Murayama, I., Inagaki, S. Takenaka, M., Nagato, M., Yamamoto, T., Nakayama, H., Yamaguchi, E. (2002). CSCL Design Experiments in Japanese Elementary Science Education: Hypothesis Testing Lesson and Collaborative Construction Lesson. *Paper presented at the Annual Meeting of the AERA, New Orleans*.

Oshima, J., & Oshima, R. (2002). Coordination of Asynchronous and Synchronous Communication. In Koschmann, T., Hall, R., Miyake, N. (Eds.), *CSCL2. Mahwah, NJ: Lawrence Erlbaum*.

Scardamalia, M. (2002). Collective cognitive responsibility for the advancement of knowledge. In B. Jones (Ed.), *Leberal Education in the Knowledge Age*. Chicago, IL: Open Court.

Collaborative E-Test Construction

Using Predicted Response-Time and Score Distributions to Improve Reliability

Pokpong Songmuang and Maomi Ueno
The University of Electro-Communications, Japan

Abstract: Analysis of collaborative e-test construction identified the number of test-authors as the most important factor in test validity, while test reliability depends more on participation of an expert. Based on these findings, a collaborative e-test construction system was developed that uses predicted response-time and score distributions to improve the reliability of tests constructed by novice test-authors. A gamma distribution is used as the predicted response-time distribution, and a mixed model of binomial distributions is used as the predicted score distribution. An experiment in which a novice and an expert test-author each constructed tests by using and not using these predicted distributions showed that those constructed using them were more reliable, although those constructed by the expert had even higher reliability.

Keywords: Collaborative e-test construction, reliability, predicted response-time distribution, predicted score distribution.

1. INTRODUCTION

Test administration on computers has become more common over the past decade. This "computer-based testing (CBT)" is done using either tests developed specifically for the computer or tests converted into a computer-based format. More recently, along with the diffusion of e-learning, CBT has been extended to web-based testing, or "e-testing." This e-testing has become a common method of evaluation for e-learning, and much attention has been paid to the use of e-testing to deliver an on-line test function to various places. Moreover, it enables collaborative test construction by several test-authors in different places. There are many advantages to such collaboration.

- It provides validation-checking mechanisms as part of the criticism process, something a machine does not provide (Miyake 1986).

Please use the following format when citing this chapter:

Songmuang, P. and Ueno, M., 2007, in IFIP International Federation for Information Processing, Volume 230, Knowledge Management for Educational Innovation, eds. Tatnall, A., Okamoto, T., Visscher, A., (Boston: Springer), pp. 195–201.

- The distributed cognition provides an opportunity to distribute work activities, thereby improving the complex information analysis (as described, for example, by Hutchins and Klausen 1996).
- It enables effective and efficient solving of ill-structured problems (Simon 1973), which need complex expert knowledge to solve.

By applying collaboration to e-test construction, we can obtain several benefits, including stimulation of test-authors' reflections and thus improved test validation, increased test reliability due to distributed cognition; and more sophisticated test construction due to the sharing of expert knowledge, particularly tacit knowledge and ill-structured knowledge.

In a previous paper, Ueno (2005) proposed a web-based computerized testing system for assisting test-authors in sharing the used item database and in collaborative test construction. However, this paper did not focus on a collaborative e-test construction system and did not provide any analysis of collaborative test construction.

Our interest here is improving the effectiveness of collaborative test construction. We compared the effectiveness of test construction by one, three, and five test-authors. The effectiveness was measured in terms of reliability and validity based on test theory (as described by Lord and Novick (1968), for example). The results showed that the reliability of a test constructed by an expert or a group of test-authors including an expert was better than that of one constructed by novice test-authors alone. They also showed that test validity increased with the number of test-authors. The main idea of this paper is to describe a collaborative e-test construction system that provides a predicted response-time distribution and a predicted score distribution that can be used to improve the reliability of tests constructed by novice test-authors. We use a gamma distribution (Ueno and Nagaoka 2005) as the predicted response-time distribution and a mixed model of several binomial distributions as the predicted score distribution. Both distributions help a test-author better understand the status of a constructed test. An experiment was performed to compare the reliability of tests constructed by an expert and by a novice test-author with and without the distributions. The reliability of those constructed using them was better although those constructed by the expert had even higher reliability.

2. TEST THEORY

Extensive research related to test construction can been summed up as a test theory (as described by Lord and Novick 1968). Traditional test theory describes two concepts related to test construction criteria.

2.1 Validity

Validity can be defined in a number of ways. in the area of test theories (For example, Lord and Novick 1968). This paper employs one of the most popular definitions. In this definition, the validity means that the ability actually measured by test item represents the ability which should be measured. In the other words, the validity indicates that the test item content exactly reflects the test domain. Content validity checking is required intuitive judgment of test-author which machine is unable to provide it.

2.2 Reliability

The central concept of classical test theory using statistics exists in the concept of "reliability." Test theory assumes that the square root of the reliability is the correlation between the true and observed scores (Lord and Novick 1968). Consequently, Cronbach's α can be used as a measure of test reliability. Recently, a more sophisticated model, item response theory (IRT), has replaced classical test theory. Here we use the test information function of IRT as the measure of test reliability (Lord and Novick 1968).

According to this theory, the validity and reliability of a test should both be maximized for it to be a good test.

3. COLLABORATIVE E-TEST CONSTRUCTION ANALYSIS

To analyze the effectiveness of collaborative test construction, we compared the validities and reliabilities of tests constructed by different numbers of test-authors (one, three, and five) with and without the participation of an expert in the test domain. The constructed tests measured Japanese language proficiency and were equivalent to the Level 4 Japanese Proficiency Test given by the Japanese government. (Level 1 is the highest, and level 4 is the lowest.) The tests were constructed based on the same item database, and data on the construction process was collected and stored.

The validity of each test was measured using a test item database we constructed including some incorrect items. The number of incorrect items included in the test was used as the measure of its validity. To evaluate test reliabilities using IRT, we used a three-parameter logistic model:

$$p_i(\theta) = c_i + \frac{(1+c_i)}{1+e^{-Da_i(\theta-b_i)}}$$

where θ is the person ability parameter, and a_i, b_i, and c_i are item parameters. The b_i represents the item location, which, in the case of attainment testing, is referred to as item difficulty. The a_i represents the discrimination of the item, that is, the degree to which the item discriminates between persons in different regions on the latent continuum. This parameter characterizes the slopes of the item response curves. For items such as multiple-choice, parameter c_i is used in an attempt to account for the effects of guessing on the probability of a correct response. Using a Bayesian method, we estimated the values of these parameters from the data for the constructed tests. The following function was used to calculate the test information which we used as an index of test reliability.

$$I(\theta) = \sum_{i=1}^{m} a_i^2 p_i(\theta)[1 - p_i(\theta)]$$

We used the Pearson correlation coefficient and t-test value to calculate the correlation between test construction parameters as shown in the table 1.

Table 1: Pearson correlation coefficients and t-test values between test construction parameters

	a	b	c	d	e	f	g	h
a								
b								
c	0.94 (0.010)	0.10 (0.010)						
d	-0.76 (0.076)	0.13 (0.578)	-0.77 (0.011)					
e	-0.33 (0.042)	0.86 (0.000)	-0.13 (0.013)	0.23 (0.001)				
f	0.97 (0.008)	-0.21 (0.037)	0.70 (0.071)	-0.63 (0.009)	-0.49 (0.015)			
g	0.88 (0.040)	-0.22 (0.037)	0.90 (0.007)	-0.86 (0.042)	-0.41 (0.099)	0.89 (0.006)		
h	0.95 (0.272)	-0.47 (0.206)	0.24 (0.010)	-0.49 (0.323)	-0.18 (0.014)	0.43 (0.008)	0.51 (0.038)	

a. number of test-authors
b. participation of an expert
c. average test construction time
d. average number of incorrect items
e. average test information
f. average number of times an item was added
g. average number of times an item was deleted
h. average number of times an item was created

- The test information and the participation of an expert were highly correlated.
- The average number of incorrect items was correlated with the test construction time and the average number of times an item was added.
- The test construction time was correlated with the number of test-authors and the average number of times an item was added.

Test validity increased with the number of test-authors and construction time, while test reliability depended on the participation of an expert.

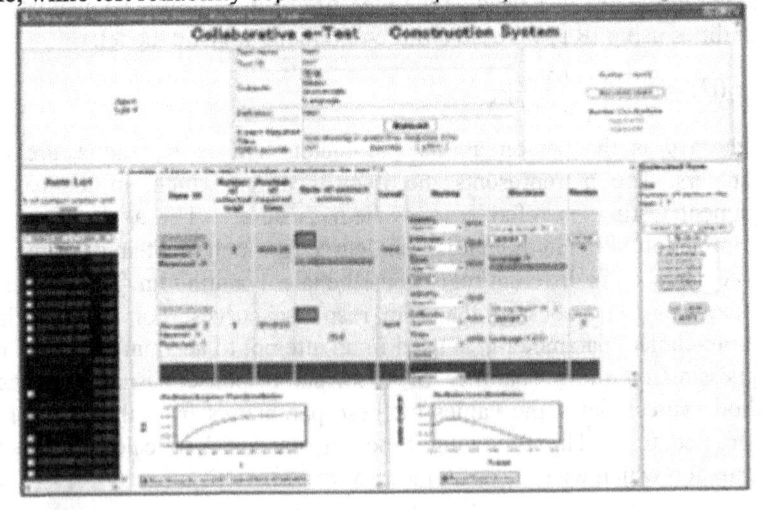

Figure 1: Collaborative e-test construction system

4. COLLABORATIVE E-TEST CONSTRUCTION SYSTEM

As shown by the results above, test reliability depends on the participant of an expert, so we investigated ways to improve the reliability of tests constructed by novice test-authors. We developed a collaborative e-test construction system, as illustrated in Figure 1. Its basic function is to enable test-authors in distant places to share items in a used item database and to create new items. The test-authors are able to add items to and delete items from the constructed test. The system also provides a discussion board to enable the test-authors to share opinions and ideas during their collaboration.

The main focus of this description is on the use of the predicted score and response-time distributions, which are used to support the authors' decision making, as shown at the bottom of Figure 1.

4.1 Predicted score distribution

The system presents the predicted score distribution of the test being constructed to enable the authors to visualize its current status. The set of the probabilities of correct answers for m items $\Theta = \{\theta_i\}, \{i = 1,...,m\}$ is estimated using historical data. A Bayesian estimation based on a binomial distribution is estimated using:

$$\theta_i = \frac{n_i + a'}{n + a'}, (i = 1,...,m), \qquad (1)$$

where m is the number of items on the test, n_i is the number of examinees who provided correct answer, n is the number of examinees, and a' is the value of a hyper parameter. We set $a' = 1$, reflecting our assumption that the prior distribution is uniform. Let $x, (0,...,m)$ be a score random variable for a test with m items. The mixed model of several binomial distributions is defined by

$$p(x|\Theta) = \sum_{i=1}^{m} [p(m_i)p(x|m_i, \theta_i)], \qquad (2)$$

where $m_i(1,...,m)$ means the *i*-th model.

4.2 Predicted response-time distribution

Ueno and Nagaoka (2005) analyzed e-learning time based on a gamma distribution with parameters α and β representing the complexity of the learned content and the expected time of a simple cognitive process.

To visualize the current status of the constructing test required time, the proposed study provides a predicted response-time distribution. We use the gamma distribution described by Ueno and Nagaoka as the predicted response-time distribution along with item historical data. We assume that any testing process consists of α repetitions of simple cognitive processes.

Moreover, the response time for a simple problem-solving process is assumed to follow a distribution so as to maximize, and, given minimum response time t_0 and average response time E, what is given by:

$$H[f_s(t)] = \int_{t_0}^{\infty} f_s(t) \log f_s(t) dt . \qquad (3)$$

The required time for a simple problem-solving process is given by an exponential distribution:

$$f_s(t) = \frac{1}{\tau} e^{-(1/\tau)t} . \qquad (4)$$

While the testing process is generally viewed as consisting of α layers of the process given by (4) and is thus obtained by a convolution integral of (4), we introduce β, the time required for solving a simple problem, and calculate α convolution integrals under the restriction that $\alpha\beta = E$. (5)

Thus, the gamma distribution obtained as the distribution model for the required learning time is

$$f(t) = \frac{t^{\alpha-1} \exp\left(-\frac{t}{\beta}\right)}{\beta^{\alpha} (\alpha - 1)!} . \qquad (6)$$

The predicted response-time distribution is then given by

$$F(t) = \begin{cases} 0 & t < t_0 \\ \int_{t_0}^{t} f(t) dt & t \geq t_0 \end{cases} . \qquad (7)$$

5. EXPERIMENT

We compared the reliability of tests constructed by a novice test-author and by an expert test-author with and without the predicted distributions. The constructed tests measured Japanese language proficiency and were equivalent to the Level 4 Japanese Proficiency Test given by the Japanese government. The tests were constructed based on the same item database, and data on the construction process was collected and stored. Each constructed test had about 30 items.

Table 2. Average information of constructed tests

Test-author	Novice w/o distributions	Expert* w/o distributions	Novice with distributions	Expert* with distributions
Test information	3.279	3.952	4.805	8.735

**Had Japanese language proficiency equal to or better than Level 2 on Japanese Language Proficiency Test.*

As shown in Table 2, the average information of the tests constructed using the predicted distributions were higher. This means that the test reliability was improved by using the predicted distributions.

However, the average information of the tests constructed by the expert with and without the distributions was higher than that of the novice. This means that expert knowledge is still an important factor in test construction.

6. CONCLUSION

We have developed a collaborative e-test construction system that provides a predicted response-time distribution and a predicted score distribution that can be used to improve test reliability. A gamma distribution is used as the predicted response-time distribution, and a mixed model of binomial distributions is used as the predicted response-time distribution.

To evaluate the effectiveness of this approach, we compared the reliability of Japanese language proficiency tests constructed by a novice test-author and by an expert test-author with and without the predicted distributions. The tests constructed using them had higher reliability although those constructed by the expert had even higher reliability. We plan to develop an agent system that plays the role of a domain expert in order to increase the reliability of tests constructed by novice test-authors.

7. REFERENCES

Hutchins, E. and Klausen, T. (1996), Distributed Cognition in an Airline Cockpit, In Middleton, D. and Engeström, Y. (eds.), *Communication and Cognition at Work*, Cambridge University Press, Cambridge, pp. 15–54.

Lord, F. M. and Novick, M. R. (1968), *Statistical theories of mental test scores*. Reading, MA: Addison-Wesley.

Miyake, N. (1986), Constructive Interaction and the Iterative Process of Understanding, *Cognitive Science*, vol. 10, no. 2, pp. 151–177.

Simon, H. A. (1973), The Structure of Ill-Structured Problems, *Artificial Intelligence*, vol. 4, pp. 181–201.

Ueno, M. (2005), Web based computerized testing system for distance education, *Educational Technology Research*, vol. 28, pp. 59–69.

Ueno, M. and Nagaoka, K. (2005), On-Line Analysis of e-Learning Time based on Gamma Distributions, *Proceedings of World Conference on Educational Multimedia, Hypermedia and Telecommunications 2005*, pp. 3629–3637. Norfolk, VA: AACE.

Three Phase Self-Reviewing System
for Algorithm and Programming Learners

Tatsuhiro Konishi, Hiroyuki Suzuki, Tomohiro Haraikawa and Yukihiro Itoh
Faculty of Informatics, Shizuoka University, Japan

Abstract: This paper introduces an electronic report submission system that helps effective learning of algorithms and programming. It proposes a three-phase reviewing system that involves self-reviewing of algorithms, self-reviewing of programs and staff reviewing. This is an improvement of our existing two-phase reviewing system that only supports the latter two phases. In the additional phase for algorithmic checking, learners describe an algorithm graphically using PAD, compile it, and execute it to verify their algorithm first without being troubled by syntax of a programming language; this supplies effectiveness to the efficient self-reviewing system.

Keywords: Algorithm, programming, learning, self-review system.

1. INTRODUCTION

This paper introduces an electronic report submission system that helps the learning of algorithms and programming. Learners' programs typically contain numerous mistakes and must be reviewed again and again before becoming acceptable. While it takes a few hours until the learners can get staff comment, the turnaround time must be shortened, so as not to distract the learner's concentration. Thus we have proposed two-phase reviewing: an automated self-reviewing phase for improving efficiency of learning, and a careful reviewing phase by staff.

Although algorithms should be represented independently of any specific programming language, present algorithm education is filled with language-dependent explanations and practices. In such a situation it is doubtful that learners can be conscious of the algorithm itself and some researchers claim that teaching of algorithms and of programming should be separated (Crews 1998). A flowchart or a PAD (Program Analysis Diagram) is used for representing algorithms and there are tools for editing and executing algorithms (Maezawa 1984, Hitachi Systems & Services), however, most of them still depend on a specific programming language. Therefore we

Please use the following format when citing this chapter:

Konishi, T., Suzuki, H., Haraikawa, T. and Itoh, Y., 2007, in IFIP International Federation for Information Processing, Volume 230, Knowledge Management for Educational Innovation, eds. Tatnall, A., Okamoto, T., Visscher, A., (Boston: Springer), pp. 203–210.

develop a language-free algorithm representing system and an algorithm validation support system, and propose a method of algorithm education using these systems.

We constructed a three-phase reviewing system that involves self-reviewing of algorithms, self-reviewing of programs, and staff reviewing, by improving our existing two-phase reviewing system that only supports the latter two phases. Through our practical operation, we learned that learners tend to be hooked on syntax when teachers let them check their code. We are now implementing a new system that also allows an algorithmic check before coding. Learners can graphically represent, compile, and execute their language-free algorithms to verify them without being troubled by syntax of a programming language; this supplies effectiveness to the efficient self-reviewing system.

In this paper, we first introduce our two-phase reviewing system. We then discuss extension of the system, especially methods of assisting algorithm education and algorithmic check.

2. TWO-PHASE REVIEWING SYSTEM

We had already adopted a self-reviewing system in programming education to reduce the turnaround time. At first, we thought that a GCC compiler was sufficient for learners to review their programs locally if its warning level is maximized. However inspection of submitted reports reveals that many typical mistakes are not detected as errors or do not receive warnings. For example, an erroneous code "**if (1 <= month <= 12) { ... }**" never receives a warning since it is considered as a condition that compares logical value of **1 <= month** (0 or 1) and an integer value **12**. The condition perfectly satisfies C syntax. Such kinds of mistakes are left uncorrected until staff notice and write a reviewing comment. The fact leads to a long turnaround time and heavy staff loads. Fortunately we know some code reliability checkers for embedded systems. For continuous fault-free operation required for embedded systems, the tools perform strict source-level analysis to point out any doubtful scraps. Some of them find meaningless conditions or operators (such as **return r++;** for a local variable **r**), make a string comparison using an operator ==, and even find typical array-index overruns. We applied such tools to the collected reports, and employed one product for self-program-reviewing in our two-phase system.

2.1 SELF REVIEWING PHASE

As we supposed, the reliability checker was useful but it was sometimes unusable for educational use due to too many suggestions or too few detected mistakes. Also, it has a user interface for professional use. So we decided to make a '*wrapper*' of our reliability checker which can both provide flexible levels of suggestions and a user-friendly interface.

We designed an email-based report submission system which sends back compilation status and source-code reliability analysis immediately. The email report should consist of the report text as an email body and sources (and headers) as attachment files. When a **recv** script receives a report via a Mail Transfer Agent (MTA), it tears the attachments off and forks to the gcc and the reliability checker in this order. Subsequently, it sends back the result by email. The wrapper is designed to suppress or replace some over warning suggestions for untrained programmers. The wrapper suppresses all strong suggestions about Y2K problems and code-optimizing directions, and replaces some suggestions with [*Info*]s. An original [*Info*] for "Line 11" in Fig.1 was "*Using a pointer for accessing to array "**month[i]**" instead could generate smaller or faster object code.*", which is from the viewpoint of the embedded tools.

A learner who receives a modified suggestion as a reply can re-submit their report, and repeat this self-reviewing process depending on their need. A learner can also browse their submission history, every automatic reply and additional reviewing comments from a staff including scores (described in the next subsection) at a web site.

[Reviewer's Comment] (handtyped)
You must verify the behavior of your program before submitting your report.
[Compilation Errors and Warnings]
None. (Congratulations!!)
[Suggestions by Reliability Checker]
Line10: Wrong || usage between "**1 <= i**" and "**i <= 12.**"
Line9: **for** statement contains wrong comparison "**i > 0.**"
Line11:"**month[i]**" overruns because 12 is assigned to "i" by a **for** statement at line9.
Line11: [Info] You may use a pointer for accessing to array "**month[i].**"
Line14: Using an increment or decrement operator to the return value "i++" makes no effect.
Line7: Variable "**month**" is not referenced.

Fig. 1: an Example of Feedback (translated)

2.2 STAFF REVIEWING PHASE

Since reports are usually refined repeatedly via a self-reviewing process, the staff only have to review their best reports; this greatly reduces the reviewers' task. The reviewing screen is separated into two panes: the first one displays scoring buttons, a comment field and a contents selector. The second one initially displays a summary composed of the compilation status, reliability analysis and body part of a report. It also shows any source file selected by a staff. Comments and scores are immediately reflected on the web site and learners can submit their reports again at this point, too.

3. THREE-PHASE REVIEWING SYSTEM

We are constructing an improved system based on a three-phase model. The self-reviewing phase of the previous model is now divided into *self-program-reviewing* and an additional phase named *self-algorithm-reviewing*. The ideal flow is given below.

The first phase: The phase helps learners to fix their algorithm. A web-based *algorithm editor* enables learners to represent algorithms independently of any specific programming language. A submitted algorithm is compiled on a server by an *algorithm compiler*. Learners can download the object code to execute and verify the algorithm. Learners can repeat this phase to make their algorithms accurate.

The second phase: This phase helps learners to verify their programs by themselves. A learner writes his/her program in C language at this phase, and then submits it with an algorithm representation created in the first phase. If the program is successfully compiled, a *correspondence checker* verifies whether the submitted program is implemented correctly in accordance with the associated algorithm. In addition to the reliability status, learners can also inspect the correspondence via a web-based *correspondence viewer*.

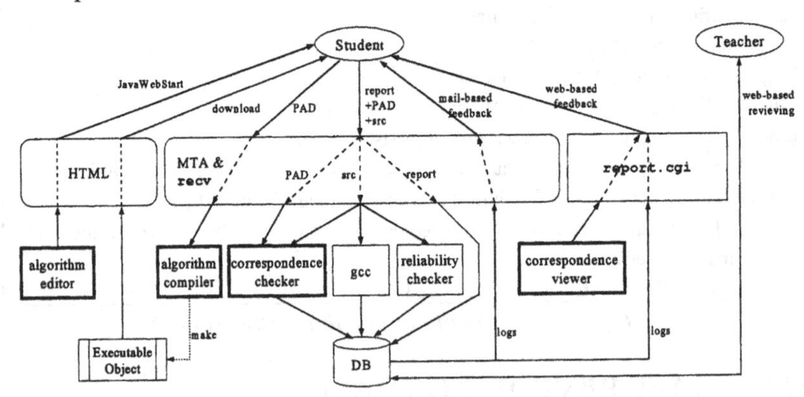

Fig. 2: Three-phase reviewing system

The third phase: This phase provides staff hand typed comments to learners. It is similar to the staff reviewing of the two-phase system, except that the phase now has a potential to provide additional information to the staff; correspondence between submitted source code and a standard algorithm written by staff.

We designed this three-phase system (Fig.2) that employs the four components mentioned above: an *algorithm editor*, an *algorithm compiler*, a *correspondence checker*, and a *correspondence viewer*, in addition to the previous two-phase system. We have constructed these components as distinct four assisting systems: a language-free algorithm representing system, a language-free algorithm validation support system, an algorithmic checker, and the two-phase reviewing system. We adopt the language-free algorithm representing system as the *algorithm editor* (Shinmura 2003). The main feature of the language-free algorithm validation support system is used for the *algorithm compiler*. The algorithm checker works as the *correspondence checker*.

3.1 SELF ALGORITHM REVIEWING

(1) Algorithm editor and Algorithm representation

Our *algorithm editor* adopts PAD representation. It has functions to help users to edit PAD expression easily. Additionally, it should be able to provide an appropriate operation set to users. In order to decide the set, we discuss representation policy for each operation. The representation of an operation has to satisfy the following requirements.

i. Learners can write algorithm by using the representation without learning any specific programming languages.

ii. The representation includes no ambiguity.

iii. The granularity of the operation should be controllable. Too large granularity of an operation allows a learner to jump into the goal using too few operations. On the other hand, if the granularity is too small, a learner can't represent his/her algorithm intuitively.

iv. It has levels of both concrete and abstract representations. One of the essential aims of algorithm education is to make learners learn how to grasp problem solving procedures in an abstract level. For example, linked lists can be represented by structures and pointers in C language. In concrete level, operations on the linked list are described by such terms as "pointer", "structure" and so on. However, learners should consider the solving process abstractly by using terms like "link", "node". So, both concrete and abstract words should be provided to describe algorithms.

Solution for i and ii: When someone describes an algorithm by any formal languages, they have to study notations and the grammar of the language. In order to avoid such extra work, we adopt natural language as a method of describing an operation. However, unrestricted natural language may be ambiguous, so we restrict the vocabulary and the sentence pattern. For learner's convenience, we prepare acceptable sentences as templates, and let learners select a template from a menu.

Solution for iii: Appropriate granularity of description depends on the goals of exercises. Therefore our system allows staff to select an appropriate granularity by selecting available templates for each exercise.

Solution for iv: In order to let learners represent algorithms abstractly, the *algorithm editor* provides templates which correspond to abstract operations to abstract data structures. We surveyed explanations of algorithm in textbooks of programming and found 7 typical data structures used to describe algorithms abstractly; list, binary tree, table, stack, heap, matrix and queue (Suzuki 2001). Based on the survey, we prepare templates to represent algorithm abstractly. When a staff member intends to let learners represent their algorithms abstractly, they select such templates as mentioned above.

Fig.3 shows a screenshot of our *algorithm editor*. In Fig.3, (1) is the area for algorithm editing. An example of an algorithm representation is displayed. (2) shows the list of variables, and (3) is the reduced drawing of (1). In the area (1), learners draw algorithm representation by mouse operation, menu selection and keyboard input.

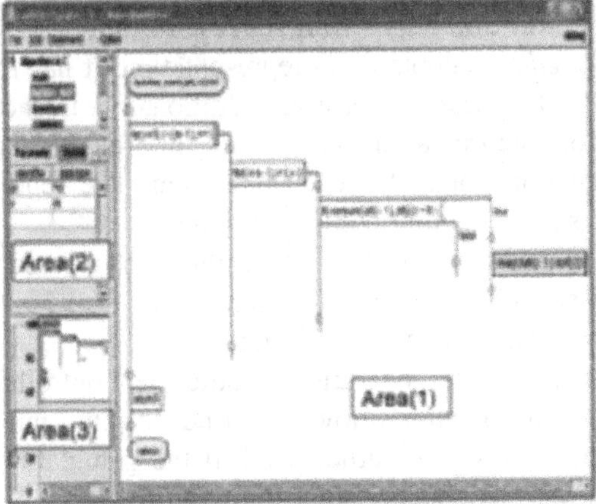

Fig.3: Screenshot of the algorithm editor

(2) Validating algorithms by learners

In order to make it possible to validate algorithms by learners, we have to develop the function of executing represented algorithms (*algorithm compiler*). In our previous work, we constructed a system which converts abstract representations of operations into source codes in a specific programming language (Suzuki 2001). We use the system as the *algorithm compiler*. With the *algorithm editor* and the *algorithm compiler*, a learner can review their algorithms in the following way: First, a learner downloads the *algorithm editor* from the web. Next, he/she writes his/her algorithm by the editor and saves it as an algorithm file. If he/she wants to execute the algorithm, he/she submits the algorithm file to our server. Then our system compiles the algorithm, and creates an executable file. The learner can locally execute the algorithm and can also validate its behavior.

3.2 SELF PROGRAM REVIEWING

In the second phase, a learner implements the validated algorithm using a programming language, in order to acquire knowledge on syntax of the programming language and techniques on implementation.

We think that mistakes in an erroneous program can be categorized into two types. One is caused by misunderstood syntax or mistyping. The other is caused by the fact that a learner can't break down an operation in algorithm into smaller pieces, or can't convert operations into a set of statements of a programming language. Compilers and code reliability checkers can only check the former mistakes. The latter can be checked by comparing a learner's algorithm representation with his/her source code. The method of checking such correspondence is as follows (Suzuki 2001):

The *correspondence checker* breaks down an operation into the smallest grain-sized operations, which are comparable with statements of a programming language. When there are some operations represented abstractly, many possible candidates can be generated from them. The checker searches for the candidate most similar to the learner's program. Through the searching, the checker stores information of correspondence between operations in the algorithm representation and statements in the learner's program.

By using these components, a learner reviews their programs as follows: at first, a learner writes a program based on their validated algorithm and submits both the algorithm file and C program to our server. Then gcc, the code reliability checker, and the *correspondence checker* work. Diagnoses by the components are stored in a database and are immediately sent to them by email. Additionally, they can see the correspondence between their algorithm and program by using the *correspondence viewer* that works on a web browser (Fig.4). Learners can easily find operations/statements which do not correspond to the program/algorithm. In addition, when a learner places a mouse cursor on an operation/statement, the statement/operation which corresponds to it changes its color. By these functions, learners can confirm whether they correctly implement their algorithms.

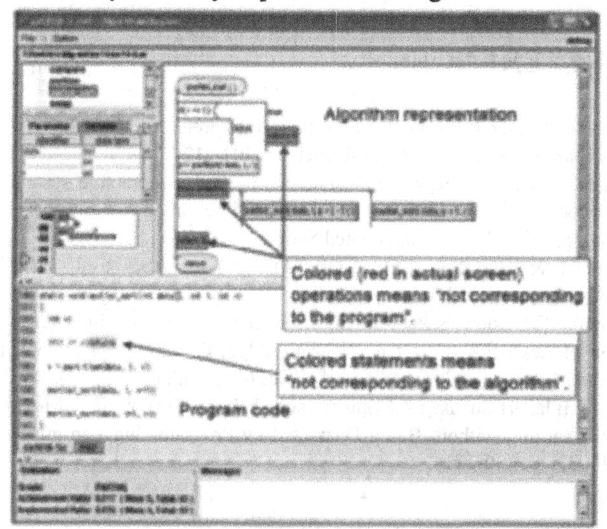

Fig.4: Screenshot of the correspondence viewer

3.3 STAFF REVIEWING

Finally, a staff member reviews programs, algorithms, and reports which are submitted by email. The staff can refer to all the diagnoses given to the learners. Moreover, they can use the *correspondence checker* and the *correspondence viewer*, in order to compare a learner's program with a standard algorithm that they write. Such usage makes it easier to find bugs which are not found by learners.

4. CONCLUSIONS

We proposed a self-reviewing system that realizes efficient and effective learning of algorithm and programming. The self-reviewing system is a front-end of our three-phase electric report reviewing system. The new first phase, *self-algorithm-reviewing*, allows learners to concentrate on representing their language-free algorithms in a PAD before writing their programs. An algorithms sent to the server is internally translated into C language and compiled. The system makes the object code downloadable by learners. Learners can repeatedly submit, validate and correct their algorithms by themselves. Learners write their codes in the second phase of *self-program-reviewing*. Formally, the phase includes not only a syntax and reliability check, but also self-correspondence-check between a learner's algorithm and his/her program. The third phase newly provides a staff with detailed analysis report; that will be of great help in scoring or writing hand typed comments. Now we are planning to apply the system to actual classes of algorithm and programming in our university.

5. REFERENCES

T. R. Crews, U. Ziegler (1998): The Flowchart Interpreter for Introductory Programming Courses. Proceedings of FIE '98 Conference, pp.307-312.

H. Maezawa, M. Kobayashi, K. Saito, Y. Futamura (1984): Interactive system for structured program production, Proceedings of the 7th international conference on Software engineering, pp.162-171. Florida, United States.

Hitachi Systems & Services, Ltd.: TOPITAL: PAD CASE tool, http://www.hitachi-system.co.jp/topital/index.html

Hiroyuki Suzuki, Takaomi Sakai, Tatsuhiro Konishi, Yukihiro Itoh (2001): Automated Evaluation of Learner's Pro-grams by using Algorithm Representations Independent of Programming Languages, Proceedings of ICCE2001, vol2, pp.883-890.

K.Shinmura, E.Iida, H.Suzuki, T.Konishi, and Y.Itoh (2003): The Method to Support Algorithm Learning without Being Distracted by Programming Lan-guages, Proceedings of ICCE2003, pp.1210-1214.